THE ATOM BESIEGED

THE ATOM BESIEGED
Extraparliamentary Dissent in France and Germany

Dorothy Nelkin

Michael Pollak

The MIT Press
Cambridge, Massachusetts, and London, England

This book was set in Fototronic CRT Baskerville
by The Colonial Cooperative Press Inc.
and printed and bound by The Murray Printing Company
in the United States of America.

Library of Congress Cataloging in Publication Data

Nelkin, Dorothy.
   The atom besieged.

   Includes bibliographical references and index.
   1. Atomic energy policy—France—Citizen participation. 2. Atomic energy policy—
Germany, West—Citizen participation. I. Pollak, Michael, joint author. II. Title.
HD9698.F72N44   333.79'24   80-15324
ISBN 0-262-14034-9

Contents

Abbreviations

AC	Autorisation de Création
AEK	Aktionsgemeinschaft der Bürgerinitiativen für Energiesicherung und Kerntechnik
AFRPNE	Association Fédérative Régionale de Protection de la Nature de l'Est
AFU	Arbeitsgemeinschaft für Umweltfragen
APRI	Association de Protection contre les Rayonnements Ionisants
AUD	Aktion Unabhängiger Deutscher
BBU	Bundesverband Bürgerinitiativen Umweltschutz
BDI	Bundesverband der Deutschen Industrie
BMFT	Bundesministerium für Forschung und Technologie
CDS	Centre des Démocrates Sociaux
CDU	Christlich Demokratische Union
CEA	Commisariat à l'Energie Atomique
CERN	Centre Européen de Recherche Nucléaire
CFDT	Confédération Française Démocratique du Travail
CGT	Confédération Générale du Travail
CNRS	Centre National de la Recherche Scientifique
COLINE	Comité Législatif d'Information Ecologique
CRS	Compagnies Républicaines de Sécurité
CSFR	Comité de Sauvegarde de Fessenheim et de la Pleine du Rhin
CSU	Christlich Soziale Union
DATAR	Délégation à l'Aménagement du Territoire et à l'Action Régionale
DGB	Deutscher Gewerkschaftsbund
DUP	Déclaration d'Utilité Publique
DWK	Deutsche Gesellschaft für Wiederaufarbeitung von Kernbrennstoffen
EDF	Electricité de France
EEC	European Economic Commission
FDP	Freie Demokratische Partei
FEN	Fédération de l'Education Nationale
FFC	Fédération Française des Consommateurs
FFSPN	Fédération Française des Sociétés de Protection de la Nature
FRAPNA	Fédération Rhône-Alpes de Protection de la Nature
GAZ	Grüne Aktion Zukunft
GECENAL	Groupe d'Etude et de Concertation pour l'Environnement et la Nature en Alsace
GSIEN	Groupe de Scientifiques pour l'Information sur l'Energie Nucléaire
IAEA	International Atomic Energy Agency
IEJE	Institut Economique et Juridique de l'Energie
IEO	Immediate Effect Order
JUSO	Jungsozialisten in der SPD
KWU	Kraftwerk Union A.G.
MRG	Mouvement des Radicaux de Gauche

NPD	National Demokratische Partei Deutschlands
NWK	Nordwestdeutsche Kraftwerke A.G.
PCF	Parti Communiste Français
PEON	Commission Consultative pour la Production d'Electricité d'Origine Nucléaire
PR	Parti Républicain
PS	Parti Socialiste
PSU	Parti Socialiste Unifié
RPR	Rassemblement pour la République
RSK	Reaktor Sicherheits Kommission
SCSIN	Service de Sécurité des Installations Nucléaires
SDS	Sozialistischer Deutscher Studentenbund
SEPANSO	Société pour l'Etude, la Protection, et l'Aménagement dans le Sud-Ouest
SFDE	Société Française du Droit de l'Environnement
SFIO	Section Française de l'Internationale Ouvrière
SIRAE	Service d'Information et de Relation avec les Associations de l'Environnement
SPD	Sozialdemokratische Partei Deutschlands
SSU	Strahlenschutz Kommission
TÜV	Technischer Überwachungsverein
UDF	Union pour la Démocratie Française
WISE	Worldwide Information Service for Energy
WSL	Weltbund zum Schutz des Lebens

Preface

This book originated in our common interest in political movements that have developed to challenge science and technology and their policy effect. Of the many technological disputes over the past decade the nuclear debate has been the most virulent, long lasting, and significant. Opposition to nuclear energy technology has threatened government policy with crucial political, economic, and social ramifications.

As a highly polarized and dramatic issue the nuclear debate has generated a vast, almost entirely partisan, literature. Advocates intent on forcefully arguing either for or against nuclear power usually fail to examine their own assumptions or the possible implications of their arguments and proposals. By doing so, they often further mystify the political and social relationships underlying the dispute. Insofar as possible our analysis steps back from a spontaneistic approach to look critically at the various dimensions of this persistent controversy. We hope to clarify some of the ideological and political issues that have prolonged the controversy and assess its possible significance.

Comparing the French and the German situations seemed to be particularly useful. At the beginning of the 1970s both countries had a similar nuclear policy, provoking in each case a comparable massive antinuclear movement and mobilizing capacity. But whereas the movement had virtually no policy effect in France, it imposed a moratorium in Germany. To explain this difference, we adopted a contextual approach with a very broad focus. Institutional factors, administrative procedures, and historical experiences shape the strategic play of the protagonists in social conflicts. These factors differ from one country to another, and they have to be taken into account in an analysis. This creates a dilemma for the researcher. We try to maintain both a breadth of focus and analytical depth in each of the many issues that form the nuclear debate. We are aware that in several chapters our analysis is but a first step, merely an invitation for further detailed research.

By necessity, our approach was eclectic. Using a wide range of research techniques, we interviewed officials and activists, participated in electoral meetings and internal discussions of antinuclear groups, collected a large number of activist brochures and books as well as official documents, and examined available statistics and survey data. The combination of inside knowledge of the two societies studied and our ethnographic material provided the necessary insight to pursue our analysis.

In interpreting the historical importance of the antinuclear movement, our reading of American, French, and German sociologists and political scientists made us aware that most macrosociological theorizing is deeply rooted in cultural and ideological traditions. After long discussions we decided to avoid a global interpretation. Nevertheless, we hope we complied with C. Wright Mills's definition of method and theory: "Methods are simply ways of asking and answering questions, with some assurance that the answers are more or less durable. Theory is simply paying close attention to the words one uses. What method and theory properly amount to is clarity of conception and ingenuity of procedure, and most important in Sociology just now, the release rather than the restriction of the sociological imagination."

Acknowledgments

Many people cooperated in the research and writing of this volume by providing documents, participating in long interviews, and giving us useful criticism. We would like to thank especially those who helped us with the field research. They included many activists in the antinuclear and environmental movement, government officials, utility and industrial administrators, journalists, lawyers, scholars, and students. Heinrich Siegmann, Francis Fagnani, Jürgen Lottmann of the German Federal Advisory Council for the Environment, and the information service of Electricité de France provided invaluable research material.

The book was written at the Cornell University Program on Science, Technology, and Society. Several Cornell students were extremely helpful in organizing the research material. Jean Loup Chappelet worked on the material on scientists, Madelaine Baudinet on Creys-Malville, and Lauren Stefanelli on the courts. Throughout the entire preparation of the manuscript our research assistant, Rebecca Logan, provided invaluable criticism, checking our logic as well as our language.

Drafts of the manuscript were reviewed by Irvin Bupp, Steven Del Sesto, Francis Fagnani, Jeanne Fagnani, Bernard Feld, Sheila Jasanoff, Allan Mazur, Jean Louis Missika, Robert Mitchell, Jean Paul Moatti, Mark Nelkin, Claus Offe, Otthein Rammstedt, Jean Jacques Salomon, and Dominique Wolton. The authors are responsible for the translation of all quotes in the text.

Finally, we wish to acknowledge the German Marshall Fund which provided generous support for the research and writing of this book.

Demonstration in Brittany (Sygma Photo Service)

Gorleben demonstration (Courtesy of German Information Center)

THE ATOM BESIEGED

1 Nuclear Power and Its Critics

"Pennsylvania is everywhere." So chanted European environmentalists after the Three Mile Island accident of April 1979 as they demonstrated to stop their own nuclear power programs. Civilian nuclear power has become one of the most controversial issues in western Europe. The antinuclear movement has engaged a broad array of different groups with a wide range of social and political concerns. Their persistent opposition to nuclear power reflects more than just the fear of risk. The European movement has focused on the social and political properties of this technology—its effect on the forms of authority and power, the concepts of freedom and order, the distribution of political and economic resources, the very fabric of political life.[1] For the promoters of the technology control of the atom represents a solution to the energy problem and an assurance of future well-being. However, for those who oppose this major technology nuclear power implies a kind of society with intolerable economic and political relationships. Indeed the nuclear establishment and its administrative apparatus have come to represent the social tensions and political contradictions of a technological age.

In western Europe the major expansion of nuclear power began in the early 1970s and sharply rose after the OPEC price increases of 1973. As nuclear power became an important priority, it also generated a significant protest movement. Nuclear critics have attacked national nuclear policies and obstructed their implementation at specific power plant construction sites.

In part public concerns about nuclear power follow from a set of technical problems unique to this technology: the low-level radiation released during normal operation of a power plant, the unlikely but catastrophic possibility of a large-scale accident, the routine environmental effects of heated effluents, the problems of radioactive waste disposal, and the potential military use of the plutonium produced as a by-product of reactor operation. But these practical questions of risk form only one dimension of the nuclear debate.

Nuclear power also conveys powerful apocalyptic images of extinction. It evokes a fear made vivid by the image of Hiroshima and Nagasaki—a fear not susceptible to measurement by comparing risks and benefits.

Beyond this background of fear nuclear power symbolizes the major problems of advanced industrial society: the effect of technological change on traditional values, the gradual industrialization of rural areas, the concentration of economic activities, the centralization of decision-making power, and the pervasive intrusion of government bureaucracies. For many critics of nuclear power the development of this technology is an important example of such problems; they talk less of nuclear energy than of a nuclear society. The passion underlying the debate, the ability to mobilize a broad array of different groups to oppose government nuclear programs follows from the association of this advanced technology with such ubiquitous social and political concerns. Indeed, while much of the debate continues to dwell on technical issues of safety, the challenge to nuclear power has assumed the character of a moral crusade.

We analyze the opposition to nuclear power in France and Germany as a social movement that embodies this broad range of issues. Integrating the movement is a widely shared concern about the political and administrative relationships and the collusion of economic and political power associated with the technology. We see the widespread and persistent protests against nuclear policy as a struggle for change in the organization of power and influence; the issue is not only one of safety but of political rights and obligations. It takes place in the context of other extraparliamentary and emancipatory movements—ecologists, feminists, regional autonomists—that have emerged simultaneously during the 1960s and early 1970s with their persistent challenges to the prevailing political and social order.

Our comparison concentrates on the two leading nuclear producers of western Europe, France and Germany. France developed her civilian nuclear power program in the late 1950s based on an independent gas-graphite technology. To expand the nuclear program, the government decided in 1969 to shift to the Westinghouse light-water reactor model and produce 8,000 Mw of nuclear energy by 1976. Then after the oil crisis of 1973 the government announced a dramatic increase in planned nuclear capacity, projecting thirteen new plants to be completed by 1980 and no less than fifty reactors in twenty locations by 1985.

These major decisions produced no substantive debate in either the parliament or the parties. Nuclear power, considered an implementation of general energy and military policy, was never discussed as a distinct political issue. Even the major expansion of 1974 was announced as a fait accompli, leaving no opportunity for meaningful parliamentary debate.

Instead nuclear power became the focus of intense extraparliamentary opposition, stimulating an extraordinary proliferation of citizens' groups and political action committees and provoking large-scale, dramatic demonstrations. In 1971 French plans to locate the first light-water reactor power plant in Bugey brought out 15,000 demonstrators. This was but the first of a series of mass protests organized at nearly every projected nuclear site until the massive demonstration at the Super Phoenix breeder reactor in Creys-Malville in 1977 culminated in violence.

The German nuclear strategy evolved toward the end of the 1950s. A latecomer in the nuclear competition, Germany favored immediate construction of Westinghouse light-water reactors and long-term development of high-temperature and fast-breeder technologies. The growth of the nuclear sector began slowly, but pressure from the major electricity-consuming industries and the nuclear supplier firms converged with the increasing price of oil to push the nuclear option during the 1970s. In 1973 the fourth government atom program provided for major expansion, guaranteeing substantial government support to the nuclear industry in the hope of a five-fold increase in nuclear-generated electricity by 1985. As in France these decisions engaged a tight social network of government, scientific, and industrial interests. Despite the economic and political importance of the decisions (Germany spent 17.5 billion DM for nuclear R & D between 1956 and 1976), legislators passively endorsed the administrative proposals.

Public opposition, however, has been extremely active in Germany since 1972. Environmentalists first raised objections about the nuclear program planned for the French and German banks of the Rhine. A successful court action stopped the construction of the Wyhl plant in 1975, encouraging the organization of antinuclear groups all over the country. In 1976 and 1977 actions in Brokdorf, Grohnde, and Kalkar mobilized many thousands of demonstrators. These actions involved violent confrontations with the police but also provoked court decisions that have essentially immobilized the nuclear power program.

The details of the conflicts in France and Germany reflect power relationships and critical points of tension in the two societies. Hidden strains, contradictions, and disaffections are exposed, as conflict forces people to take sides. In the course of conflict protagonists express their beliefs, attitudes, and visions of a good society. The structure of controversy, the social class alignments, and polarizations reveal diverse agendas, priorities, and concerns. The dynamics of conflict highlight political contradictions, as critics seek channels to influence policies and governments try to reinforce their authority and legitimacy.

We have chosen to focus our study of extraparliamentary conflict on the French and German nuclear debate for several reasons. France and Germany are today the leading nations in western Europe and are comparable in terms of economic and social development. As advanced industrial societies with progressively converging socioprofessional structures, they share common economic, social, and technological problems, as well as political contradictions uniquely expressed in the nuclear debate. Both have important national nuclear industries that are major competitors on the European and world markets. While the financial backbone of these industries rests on domestic demands, a setback in nuclear expansion due to internal dissent can have broad international consequences. Thus we view the outcome of the nuclear debate in these two countries as one key to the probable future of nuclear energy in western Europe.

The antinuclear movement in France and Germany also provides a means to study in a comparative framework a number of increasingly important political questions, as governments face complex and controversial policy choices in technical areas.

To what extent does the technical nature of policy choices drive the political process? The imperatives of nuclear technology, its high capital costs, and the industrial-governmental relationships necessary to produce and control nuclear power appear to encourage similar administrative centralization regardless of the political context. But France and Germany have had very different historical experiences, reflected in their different political systems and unique cultural and administrative styles. Comparative analysis provides an opportunity to test the logic of a technological issue against the political and institutional constraints of two different cultures.

What are the structural conditions underlying the antinuclear movement? Patterns of economic growth and industrialization have changed the social structure and with it popular expectations about the nature of a good society, especially about the importance of political choice. At the same time the commitment to nuclear power and other large-scale technologies has imposed significant changes in certain political and administrative relationships. These changes are key targets for nuclear critics whose concerns are expressed in the themes and ideologies of the antinuclear movement.

How do critics express their reaction against a major technology to which their government is fully committed? In both countries organized groups have challenged nuclear policy, its implementation, and the existing policy-making and regulatory procedures. But social movements do not develop in a political vacuum. Comparing the movements in two political systems reveals differences in the dynamics of conflict, as social class alignments and cultural characteristics shape citizens' expectations and demands on political procedures. How critics actually confront their government depends on legal and administrative arrangements, available participatory channels, and anticipated government reactions. Thus the emergence and development of the antinuclear movement, its constituency, its access to expertise, its forms of expression, and its relationship to existing political institutions can be expected to reflect the social and political context.

How do different governments come to terms with challenges to technological policies that involve major economic commitments? Nuclear technology has brought government into increasingly significant partnership with industrial interests, but this very collusion of economic and political power is also a major source of public mistrust. The nuclear opposition challenges not only a technology but also political legitimacy. Governments are constrained by their need to both maintain social order and convey an image of democratic decision making. Pressed between constraints imposed by alliance with industrial interests and demands for public involvement and local control from an ever-critical public, governments oscillate between declarations of democratic goodwill expressed in participatory mechanisms and repression of nuclear protest. The amplitude of this oscillation, however, differs in the two political contexts.

Finally, what is the significance of the persistent opposition to technology, in particular to nuclear power? Is it ephemeral and likely to pass with little long-term effect, a visceral reaction against modernizing tendencies? Or can it be considered a significant movement with a vision of a future social order and a promise of basic structural change?

In France it is often suggested that the opposition is simply a result of public ignorance and inadequate information. In Germany the movement is often attributed to anarchistic groups—to a radical fringe manipulating public fear for its own political ends. We approach the opposition to nuclear power as a far more complex phenomenon: as a social movement with a broad popular base and an ideology that challenges established political institutions. In both countries the antinuclear movement is sustained by sympathetic public attitudes from a heterogeneous class constituency. It comprises a network of committed associations and scientific activists, and its themes convey a critique of the state that has captured the support of diverse groups. The population density in Europe, the scarcity of unpopulated land, and the relatively large number of nuclear power sites are compelling factors in shaping public attitudes.

Despite a flourishing literature on social movements, no consistent or standard concept has emerged. Most definitions are either too broad or too restricted to be analytically useful. According to the *Encyclopedia of the Social Sciences* the term social movement denotes "a wide variety of collective attempts to bring about a change in certain social institutions or to create an entirely new order." [2] Such broad definitions encompass an enormous range of phenomena and often fail to distinguish social movements from protest movements or pressure groups. More restrictive definitions, focusing on specific organizations or ideologies, tend to miss the diversity and richness of social movements intrinsic to their importance and broad appeal.

In describing social movements European sociologists often maintain a theoretical perspective that limits the concept to movements that produce disruptive changes in the social structure and power relationships in a society.[3] Our analysis of the antinuclear movement will suggest its political implications but cannot assess its long-term historical significance. Rather we base our analysis on the reflections of a German sociologist, O. Rammstedt, who has developed an evolutionary interpretation of social movements, defining them as a social process that unfolds in a situational context.[4] Protest groups emerge as a result of structural changes in a society, and as a movement develops, its organization, ideologies, and tactics adapt to the environmental and political circumstances in which it evolves. In the course of broadening its constituency and developing its strategies, a social movement also becomes a source of further structural change, though often in directions that may be neither anticipated nor intended.

We approach this study with a distinct political bias—that governmental capacity to tolerate radical protest and social conflict is a criterion for public freedom. The concentration and collusion of economic and political power intrinsic to the nuclear sector, and the sense of economic and technological imperative that pervades the political climate of energy decisions, have at times brought administrative or even police repression of criticism. We find this unreasonable, for the importance of the antinuclear movement lies less in its specific impact on public policy than in its ability to reveal that alternative evolutions may exist. To take such movements seriously is simply to maintain confidence in the human capacity to influence history and prevent disasters that often appear to be the immutable consequence of iron laws.

I THE POLITICAL CONTEXT OF THE NUCLEAR CONTROVERSY

2 The Nucleocrats and the Organization and Ideology of the Nuclear Establishment

Images of the nuclear establishment are shaped by different political perspectives. To the promoters of nuclear power the social network that comprises government, administration, private industry, and scientists is a pragmatic and technical enterprise devoted to the practical implementation of a policy unquestionably in the public interest at a time of declining energy resources.

To the antinuclear activists, however, the so-called nucleocrats are an advanced technocratic power elite, representing the nucleus of a future technology-based fascism. A German best seller written by Robert Jungk refers to a closely controlled *Atomstaat*.[1] French ecologists talk of *électrofascism*.[2] Indeed critics perceive the nuclear establishment as an ideological enterprise engaged in activities having profound implications for the structure of power in their societies.

Several characteristics of the nuclear enterprise in France and Germany nourish such perceptions. In each country the development of nuclear power has followed different patterns that reflect historical circumstances and different structures of public control. But in both cases the development of a nuclear power capacity requires a concentration of effort that tends to blur the distinction between public and private interests and between political and administrative choice. The nuclear establishment justifies this effort because it equates the technology with national independence and economic progress. It also embodies its nuclear plans in a set of futuristic and utopian myths.

The Evolution of a Government Industry Monopoly

After World War II France's first postwar government, formed by the returning Gaullists and the resistance movement with its significant socialist and communist influence, implemented nationalization programs in key industrial sectors. The subsequent importance of this nationalized sector and the early ties between the civil and military aspects of nuclear power in France established from the beginning a pattern of strong central government control over nuclear policy.[3] This control remains to this day unfettered by legislation dealing specifically with nuclear technology.

In 1945 the Commissariat à l'Energie Atomique (CEA) was set up as the center for nuclear research and development. As nuclear energy became commercialized in the late 1960s, Electricité de France (EDF), the state-owned utility, assumed increasing control. These two administrations have determined government policy in the nuclear field.

In 1957 the government established the Commission Consultative pour la Production d'Electricité d'Origine Nucléaire (PEON) to coordinate the activities of the CEA, EDF, and the related industries. The members of PEON represent EDF, CEA, the ministry of industry, the ministry of planning, the ministry of finance, and the directors of the major industries involved (Creusot-Loire, Pechiney-Ugine-Kuhlman, Compagnie Générale d'Electricité, Alsthom, CEM). Together they initiated the civilian nuclear energy program. By 1966 five prototypes of natural uranium gas-graphite reactors were put into production at Marcoule and Chinon. Three more units were ordered by EDF at this time; two in St. Laurent-des-Eaux and one in Bugey. These first power plants had limited generating capacity: Chinon I (1963) 70 Mw and Chinon II (1965) 210 Mw.

To EDF the French process appeared increasingly problematic in terms of its costs and limitations for future international export. Thus, just when CEA was trying to commercialize the production of gas-graphite reactors, EDF sought to innovate.

Various models were explored during the late 1960s—an enriched uranium light-water reactor at Chooz, a heavy-water reactor at Brennelis, and the fast breeder. By 1968, stimulated by market considerations and active American sales to other European countries, EDF proposed adopting the American model light-water reactor, with the intention of developing their own export capacity. CEA, however, remained committed to the French process, as did the private industrial manufacturing companies committed to the gas-graphite technology.[4]

The growing tension between CEA and EDF and the decrease of oil prices during the 1960s delayed the nuclear program. But the temptations of the export market were growing, and the Westinghouse reactor was clearly beginning to dominate the European market. Once President De Gaulle left office at the end of the 1960s, a ministerial council led by President Pompidou supported EDF's proposal to abandon the gas-graphite reactor program in favor of the light-water technology. This decision limited CEA's influence, and subsequently French civilian nuclear policy has been elaborated mainly by the planning services of EDF and several large industrial firms.

The five-year plan developed in 1970 included a program to produce 8,000 Mw of nuclear energy by 1976, with new plants at Fessenheim in Alsace and at Bugey. EDF awarded 51 percent of the new contracts to Framatome—a subsidiary of the large French steel company Creusot-Loire—and 45 percent to Westinghouse. The plans proceeded slowly, delayed in part by technical problems.[5] But pressures increased with the oil crisis in 1973 to 1974. In 1954 internal resources had provided 64.6 percent of France's energy needs; by 1974 all but 23.8 percent of her energy needs had to be imported. The situation became especially problematic with the loss of control over Algerian petroleum deposits after decolonization. When the price of petroleum quadrupled after 1973, France's trade surplus of $1.4 billion changed to a deficit of $3.5 billion.[6] Meanwhile forecasts estimated a doubling of electricity consumption every ten years.

In 1974 the government announced a plan to expand the nuclear program, projecting thirteen nuclear plants of 1,000 Mw each to be completed by 1980. The long-term plan was to build by 1985 fifty reactors in twenty locations providing 25 percent of France's energy and by the year 2000 two hundred reactors in forty nuclear parks providing more than half of France's projected energy needs. Coordinating with EURODIF, an enriched uranium plant would be built in the Tricastin region to assure an independent supply of uranium. Finally, it was planned to continue rapid development of the breeder reactor program. The Phoenix reactor, a 230 Mw demonstration model, was put into operation on July 14, 1974, in Marcoule, and a year later construction began on the 1,200 Mw Super Phoenix in Creys-Malville, intended to be operative by 1982.

Initially EDF tried to use a variety of suppliers for nuclear plants and components, but rapid expansion of the nuclear program and the pressures of international competition led to increasing concentration. In 1975 the French government administration became the major stockholder of Framatome: CEA taking a 30 percent share, Creusot-Loire 51 percent, Westinghouse 15 percent. Thus EDF works with one supplier which as a monopoly is in an excellent bargaining position to impose its conditions.

EDF is one of France's largest companies, employing over 120,000 people and investing over $2 billion each year. The long-term debts and loans necessary for continuing the nuclear program corresponded to 65 percent of EDF's real estate values. This level of investment required capital from the national and international market and inevitably limited the degree of significant public control over EDF's strategies.[7] To the left, viewing EDF's nationalized status as an important structural reform in a capitalist country, its financial policies represent an erosion of a national enterprise by private capital. EDF's contracting policies, favoring economic concentration, represent a monopoly situation in which the government, through stockholding, subsidizes a private corporation, Creusot-Loire. Besides Framatome, Creusot-Loire controls Novatome, formed in 1977 to develop the fast-breeder program. Creusot-Loire is also directly linked to the suppliers of plant components through Empain-Schneider, the third largest financing group in France.[8]

While in France private influence over the administration's nuclear policy developed only over the last ten years, in Germany private industry has always played a central role. Germany had no nuclear program during the early 1950s because the building of a German nuclear industry was forbidden by the Allied forces until 1955. Only at the end of its status as an occupied nation was the Federal Republic even allowed to do research and development in this field, and by then there were several international efforts to control nuclear energy and development. Thus Germany had to fill a technological gap in a situation of competitive disadvantage.

Germany's strategy followed a different pattern from that in France where need to develop both a civilian and military program largely explained the choice of the gas-graphite line. Germany, politically unable to build a military nuclear program, saw less need to develop her own technology. The absence of a military goal also limited the involvement of the public sector.

The importance of the private rather than the public sector was reinforced by historical circumstances. The role of German trusts in the rise of the Third Reich had sustained socialist suspicions about big corporations; even the first programs of the conservative Christian Democratic party (CDU) proclaimed the need of government control over the economy. But the division of the country in 1949 and the cold war eroded this political view. Market economy and free enterprise became the official economic ideology of the Federal Republic, constantly reinforced by comparison with the socialist part of the country.

Thus development of nuclear power relied heavily on private industry, and as a result German nuclear policy from the start was oriented to private economic interests, considerations of profitability, and future export possibilities. The strategy was to use foreign help—information, raw materials, and technology—to build up a nuclear industry able to meet national needs and later to conquer export markets. This required close cooperation with the United States.

The organization of the governmental nuclear administration reflects this strategic choice. The ministry of atomic energy and the German atomic commission were established in 1956. The members of the commission included eight scientists (among them Otto Hahn and Werner Heisenberg), two representatives of the administration (one of them the minister himself), thirteen representatives from large industrial corporations, two from financial institutions, and two from the labor unions. This membership reflects two objectives: to promote harmony and collaboration among industrial sectors involved in the development and use of nuclear energy and to develop a strong program by bringing together key German experts having important international contacts.

This early establishment of nuclear policy coincided with a growing fear of an energy crisis: internal energy production during the 1950s seemed incapable of keeping pace with the rapidly increasing needs of expanding industrialization. For the first time German industry looked outside to import coal, in particular from the United States. Nuclear power seemed to provide a solution to the energy problem. Then the drop in oil prices in the early 1960s removed the imperative for this policy option.[9]

In 1957 the Eltviller program had proposed the production of five prototype plants that would test different technologies. This program was technologically too ambitious, and the first atom program (1956 to 1962) based on this plan finally favored the immediate construction of light-water reactors and a longer-term development of high-temperature and fast-breeder technology.

The second atom program (1963 to 1967) sustained the initial policy choices, as did the third (1968 to 1972). The first commercial reactor, Kahl on Main, was in operation in 1961, but economic constraints delayed further development: nuclear was about 20 percent more costly than traditional technologies. However, as in France the oil crisis of 1973 brought renewed interest in nuclear power.

Just a few months before the 1973 increase of oil prices, the government proposed a new energy program intended to reduce oil consumption by expanding nuclear power and maintaining internal coal production. These options were confirmed and strengthened by the oil crisis. A revised version of the energy program proposed to reduce the annual increase in energy consumption from 6.1 (annual average 1960 to 1973) to 3.8 percent, to reduce the use of oil from 55 percent in 1973 to 44 percent of overall energy consumption in 1985, and to increase nuclear energy from 1 to 15 percent. By 1985 fifty nuclear power plants were to produce about

50,000 Mw, as compared to the 2,300 Mw produced by eight nuclear power plants in 1974.[10] This fourth atom program underlined the intention to make Germany one of the first nuclear powers in the world.

The chemical industry (especially Hoechst), one of the largest electricity consumers, and the electronics industry (Siemens and AEG) played a major role in promoting the nuclear option.[11] Their influence was reinforced by several private lobbying organizations: the German Society for Nuclear Energy (Deutsche Gesellschaft für Atomenergie), composed of businessmen, politicians, and members of parliament, and the influential Association of German Engineers (Verband Deutscher Ingenieure) and its subgroup, the Association for Nuclear Technology (Arbeitsgemeinschaft für Kerntechnik). In 1959 a private nonprofit organization, the German Atomic Forum (Deutsches Atomforum) formed to coordinate promotional and public information efforts. This association, similar to the American Atomic Industrial Forum, includes almost everybody involved in the nuclear field; the 1977 membership enumerates no less than four hundred individual members, including scientists, politicians, labor unionists, and industrialists, as well as over one hundred firms and public agencies. Its activities are financed largely by grants from private industry and banks. As a nonprofit general interest organization it is tax exempt and receives public financial support.[12]

This forum represents a broad professional and political spectrum, and in the early days of nuclear power development its expertise had helped to neutralize the criticism from those concerned about the development of nuclear weapons. But its image of neutrality also masks the important political function of this organization—it serves as an informal center for scientists, politicians, and managers to negotiate major policy decisions. Indeed ecologists often refer to the forum as the atom mafia.

As in France the nuclear industry's investment needs led to rapid industrial concentration. Two corporations, Siemens and AEG, have been most important. Siemens contracted for a Westinghouse license which expired in 1970; AEG worked with a General Electric license. Together in 1969 Siemens and AEG created a joint enterprise, Kraftwerk Union (KWU), to build nuclear power plants, and in 1977 Siemens bought out the AEG share. Through KWU Siemens also controls INTERATOM, engaged in the development of experimental plants. Through INTERATOM Siemens participates in the development of the fast-breeder reactor in Kalkar (INTERATOM holds 70 percent; the Dutch Neeratom, and the

Belgian Belgonucléaire hold the rest of the stock of the Kalkar Development Corporation). Siemens, through its relations with KWU, also controls the firm developing high-temperature nuclear technology. Besides
the market leader Siemens, Babcock-Brown Boverie (74 percent controlled
by the U.S. Babcock and Wilcox), the Swiss multinational corporation
Brown, Boverie and Cie (BBC), and the U.S. firm General Atomic play a
role on the German market.

The producers of plant components are also concentrated industries,
and these are tied to the suppliers by means of capital participation. As in
France the economic risks of long-term, capital-intensive investments reduced competition and reinforced oligopolistic structures with price policies that are difficult to control.

Electricity generation and supply in the Federal Republic are officially
in the private sector, but public authorities are major shareholders in the
big utilities.[13] Throughout Germany there are 1,400 companies involved
in electricity generation and distribution, but in 1974 eight were responsible for 73 percent of the total electricity generated, and in each region
monopoly firms control the distribution of energy.[14] The two giants,
Rheinisch-Westfälische Elektrizitätswerke A.G. and Veba A.G., alone produced more than 50 percent of the nation's electricity; both are largely
controlled by the state through shareholding. The presence of federal and
Länder (state) representatives on the administrative boards of the major
utilities could guarantee public control of their strategies. But in fact the
energy policies planned by the Länder and federal governments are based
on the forecasts prepared by these firms. In turn in their nuclear investments the firms depend on their long-term relationships with Siemens;
even the banks that provide financial guarantees are stockholders of the
nuclear corporations.[15] Thus many interests converge to support nuclear
energy and minimize the financial risk of the investment required.

Briefly an oligopoly situation with virtually no competition prevails in
both Germany and France. In France the state is a shareholder of the supplier firm. In Germany public authorities are shareholders in the big utilities. In theory state monopoly over the energy sector in France and the
presence of public representatives on the administrative boards of utilities
in Germany guarantee public control. But the long-term investments involved in the nuclear option create an inertia that precludes policy flexibil-
ity. Those who occupy key positions in the social network of the nuclear
establishment—the government sector, the interested scientific commu-

nity, and the private firms—share a stake in defending their decisions, investments, and interests, and this maintains their commitment to expand the nuclear power program.

Opponents of nuclear energy underline these relationships between private and public interests; indeed they claim that, given this structure, government authorities cannot assure that safety precedes economic considerations, as required by the German federal law regulating the use of nuclear energy. The nuclear establishment, however, defends these relationships, claiming legitimacy on the basis of its political neutrality and expert competence.

Expertise, Consensus, and the Limits of Public Control

In France the public status of CEA and EDF was assumed to guarantee political neutrality. In the civil service tradition officials consider themselves less as managers of an enterprise than as guardians of the public interest.

The agents of CEA and EDF are not simple managers but high civil servants with an idea of what must be regarded in the general interest as opposed to private interests. They perceive themselves as responsible for an important part of French industrial policy. That is why they consider protest against their activities as protest against the state and the general interest that they represent.[16]

The history of EDF as a nationalized enterprise as well as a public service provides the organization with an image of political neutrality and a status beyond suspicion (au dessus de tout soupçon) as a selfless and objective institution. EDF's prestige is enhanced by its monopoly of expertise in energy and nuclear policy.

EDF's planning department was built up under the leadership of such eminent French planners and economists as Pierre Massé, who was minister of planning from 1959 to 1966 and president of EDF from 1965 to 1968. The present director of EDF, Pierre Boiteux, is a brilliant economist. This leadership inspired total confidence until criticized by the ecology movement. As a result the ministry of industry and the ministry of planning developed their own expertise in the field of energy policy in order to negotiate with EDF on an equal footing.

The French administrative bureaucracy is shaped by the higher educational system. Graduation from one of the elite establishments (notably the Ecole Polytechnique) quarantees not only a favored career in the administration but also an image of technical infallibility and a certain social exclusivity. French officials in different administrations, in the national-ized industrial and banking sectors and in big private corporations often share this common educational background.[17] They are trained in effi-cient management and socialized to deal with public problems as tech-nical issues. They share a similar outlook toward technology and its economic and political implications. A common education facilitates com-munication within this group and also excludes others.

The organization of the German technocracy is based less on a meri-tocracy than on a model of expert consensus assumed to guarantee high-quality implementation of government policy and establish an unques-tionable basis of authority. Within the government nuclear establishment a relatively small ministry essentially follows the propositions elaborated by the atom commission. Nuclear policy is prepared by subgroups involv-ing hundreds of experts who seek to reach informal consensus prior to public discussion. Consensus and discretion are leading principles of this efficient and orderly means of determining public policy, and they served to guide nuclear policy for more than a decade. Administrators remember with nostalgia those bygone days when

The atmosphere of meetings as well as of receptions was characterized by the sense of solidarity that predominated over those earlier years in Germany.[18]

Policy discussions were secret, and a gentleman's agreement united the members, assuring discretion concerning major decisions.

When the Social Democratic party (SPD) joined the government in the big coalition (with the CDU/CSU) in 1967, the dialogue between govern-ment and industry on the nuclear program began to change. In particular the SPD argued against the dominance of industrial and scientific interests over such important decisions and criticized the system of expert com-missions as too traditional, too hierarchical, too removed from political control. In 1969 the new social-liberal government changed the advisory system, transforming the atom ministry into a ministry of science and technology and dissolving the powerful atom commission. It also ex-panded the number of its advisory commissions and experts.

A major figure in German nuclear policy saw this moment as the rup-ture of the closed collaboration between science, politics, and industry.[19] But even in the expanded system (in 1975 the ministry for science and technology employed no less than 927 consultants) some 80 percent of the experts represented scientific and industrial interests, suggesting the con-tinued dominance of the scientific-industrial complex in the advisory system.[20] Indeed the organizational changes of 1969 hardly affected the old consensus over policy choices.

Similar consensus mechanisms have also guided the semiofficial Atomforum. As an important pressure group it tries to integrate everyone knowledgeable in the field from science, industry, and government into its technical working committees. Such is its high status that, once a majority consensus is reached, it is socially almost indecent to attack it publicly. The working committees meet each year at a reactor conference to discuss and resolve detailed, technically controversial problems. Often as many as 2,000 experts attend. But critics usually fail to appear, a fact often used to dismiss them.

In fact German R & D capacity in the nuclear field is concentrated in a few laboratories closely linked to the government nuclear program. Only in the last few years have universities and autonomous institutes developed the independent scientific capacity to nourish public debate and challenge the expert knowledge put forward by government authorities.

In both France and Germany the need of expertise is a justification for closed decision-making procedures; discussions of major policy issues take place in government advisory committees. Protected from outside criticism, such committees are an ideal setting where factions of the nuclear establishment can harmonize their divergent interests before they enter the public arena.[21]

Potential conflicts are further avoided through the circulation of top-level managers between different sectors, governmental agencies, private firms, and banks. This phenomenon is especially common in France.[22] Such circulation of personnel is less frequent in Germany where conflicts within the establishment are minimized by the complicated relations between government, industry, and banks through shareholding. Key persons represent the same institutions in several administrative boards and try to prevent contradictory developments. Such tight interrelationships, where a few individuals play a decisive role, serve to harmonize governmental and industrial interests. But in doing so, it becomes difficult to separate the responsibility for the promotion and the control of nuclear technology.

Critics argue that promotion and control of nuclear technology must be institutionally separated to assure adequate regulation. They suggest that outsiders with no vested interests should share responsibility for control. Nuclear officials, however, find it difficult to even imagine external control. They argue that the nuclear industry itself has a vested interest in the safety of the technology:

There are no people more critical about their own work, for sheer self preservation! . . . Jesus! If things were not okay, that could not be kept secret! Business would suffer the most in the case of an accident.[23]

The question of who controls becomes one of confidence in the good intentions and integrity of those who promote nuclear energy.

The Nuclear Fervor as a Surrogate for Nationalism

Just as the legitimacy of the nuclear establishment is sustained by its expertise, so its policies are supported by a legitimating ideology. This is based on a set of assumptions that equate nuclear power with national goals. Those engaged in the formation of nuclear policy in both France and Germany see themselves as working for the public interest. This accounts for their shocked reaction to criticism of nuclear policy; it is an attack against what they believe to be their contribution to national independence and prestige and to international economic competition.

National independence and international competitiveness are related concepts that reinforce each other: independence can only be achieved by reducing imports of raw materials (especially oil), increasing internal nuclear energy generation, and improving the international competitiveness of the nuclear industry. The nucleocrats take these objectives for granted and assume they are synonymous with the public interest.

But such assumptions may also preclude critical attitudes. French efforts first concentrated on the development of its own reactor line to minimize dependence on foreign-enriched uranium and American technology. When technological self-reliance decreased export possibilities, the French shifted to the Westinghouse reactor. Increasing the competitiveness of the French nuclear industry on the international market was to be a major contribution to national independence. In fact the reality of this policy change favored economic internationalization more than independence.

Used strategically to justify diverse political goals, the concept of independence often has more ideological than analytic content. For example, discussions about preserving independence from foreign resources focused on Arab oil. The oil boycott after the Yom Kippur war, more than an economic threat, was nearly a personal offense: "Will we accept bondage under Arab domination?"[24] Of less concern is the dependence created by the nuclear option—dependence on uranium raw materials and on loans from the international credit market necessary to implement the capital-intensive technology.

A further contradiction appeared in the response to the U.S. nonprolif-eration policy. Many European critics regarded the U.S. pressure to re-strict the export of nuclear technology as a means to preserve American technological hegemony in the international market. While this argument may contain some truth, it too overlooks the interrelated interests of American and European capital in the multinational companies dominat-ing the nuclear field.

The prestige associated with an advanced technology has also moti-vated the development of nuclear power. Technological advance is a sym-bol of independent national strength even when it is not profitable—witness the Concorde. In both France and Germany the fast-breeder reac-tor was a means to surpass American nuclear technology—as much a source of prestige as a solution to the energy problem. This prestige of the nuclear program has special significance in Germany, where the strength of the nuclear industry is a symbol for economic, intellectual, and political recovery after World War II. A journalist argues as follows:

There was no way to express German national feeling after the war. This would have been interpreted as a Nazi attitude. West Germans instead constructed their new national identity around economic growth and power. Nothing better symbolized this than the nuclear industry. Nuclear power is the sacred cow of a new German nationalism. If you are against it, the establishment considers you anti-German—a traitor.[25]

The concept of national independence is vague and malleable, easily employed to fulfill immediate economic and political goals and to mask the interests that define these goals. But the undifferentiated and often theatrical use of the concept also masks the fact that both countries can only choose among different forms of energy dependence. Thus the myth of independence limits full and open debate about which forms of depen-dence are politically acceptable and precludes analysis of the international implications of different alternatives.

Futuristic Visions

New technologies often embody a vision of a utopian social order. From the beginning nuclear technology evoked reference to mythological and very ambivalent images. Witness the names given to the first nuclear power plants: Minerva, the goddess of wisdom and war; Saturn, who ate his children and castrated his father; Phoenix, the mythological bird growing out of ashes.

Such images link life and death, and hope and fear; utopian fantasies inevitably appear to resolve the ambivalence. When administrative documents on unresolved technical problems propose futuristic solutions, they sometimes read like a twentieth-century Jules Verne novel. Professor Wolf Häfele, initiator of the German fast-breeder program, proposes to site fast-breeder reactors in poorly developed and underpopulated regions, like the borders of the glaciers in the Austrian Alps. He dreams about a world state, a "symbiosis between a new technology and a new social structure." [26] He compares nuclear power plants to medieval cathedrals, both powerful symbols of their age.

The risks of nuclear technology have provoked dramatic and often utopian schemes. The Italian physicist Cesare Marchetti proposes to resolve problems of risk by removing all dangerous technologies from the sociosphere. Around the year 2000, he argues, we should establish a nuclear center on the Canton Island in the southwest Pacific, with a concentration of fast-breeder reactors and reprocessing plants. The waste could be deposited in the corals. At a depth of 2,500 m the containers would be so hot they would continue to fall to 5,000 m. The energy in a form of liquid hydrogen could be shipped in containers of 300,000 tons all over the world. The notion of concentrating nuclear production on a remote island is but one extreme formulation of the common proposition to develop nuclear parks:

I imagine the solution to be nuclear parks, big industrial parks concentrating energy production and waste disposal . . . a kind of zone where one can optimize the flows—not infernal zones but ones where it is possible to minimize squandering, losses, and waste.[27]

Schemes to resolve the technical problem of waste disposal or the social problem of terrorism remove the technology in time and space, as if one can concentrate risk in far away regions. But such proposals escape from problems rather than solve them: the proposed sites, similar to those chosen for atomic bomb tests, are simply located far from those populations that could effectively oppose them.

Futuristic projections about the nuclear age contain implicit biases, evident in discussions about who should have access to this new technology. Fears about terrorism and social unrest lead to judgments of maturity; Aurelio Pecci, president of the Club of Rome, remarked that "Italy is not well behaved enough for nuclear technology." [28]

The most extreme proposals to resolve the problems of nuclear technology have a science fiction character. While science fiction fantasies are often the starting point of invention, in a debate where enormous future investments depend on the solution of difficult problems, they appear politically naive. Thus these fantasies are not widely shared among those in the nuclear establishment. Yet in their extreme form they vividly express the reliance on technological solutions as well as the desire to dominate nature and society that pervades the discourse of this most advanced technological sector.

Just as the antinuclear movement developed a body of doctrine, myths, and symbols, so too the nuclear establishment developed an ideology based on a defense of national independence, a definition of progress, and an image of the future. Convinced of the rationality of their position and officially responsible to meet energy needs, nuclear proponents have avoided open discussion of the controversial values embedded in the nuclear program.

They are doubtful that the public can deal with complex issues to make decisions in its own best interests. But this in itself contradicts the constitutional principles on which the political systems of France and Germany are built. In the nineteenth century the ruling class defended its privilege and justified the restriction of equal electoral rights by the lack of education of the people. Today the argument that experts must control decisions about complex issues may leave power unconstrained by democratic procedures. This political dilemma, a key issue for the antinuclear movement, was epitomized by the administrative implementation of nuclear policy.

3 The Administrative Implementation of Nuclear Policy

The convergence of political and industrial interests in nuclear policy tended to blur the distinctions between the public and private sectors. As a result, as the governments in France and Germany sought to implement nuclear policy, their commitments to promotion conflicted with their regulatory obligations. Moreover the expanding role of the central administrative bureaucracies infringed on local authority, and this became a major target for nuclear critics. To counter growing criticism, the governments of both nations developed procedural reforms that brought more local consultation in the licensing of nuclear power plants. We analyze these reforms in the context of the very different administrative styles of the two countries. We suggest how they have helped to politicize the bureaucracies themselves, destroying the image of neutrality that has justified expanded administrative power and increasing the points of tension through which critics of nuclear power are able to affect public policy.

The French Administrative Style of Centralized Bureaucracy

The relationship between administrative bureaucracies and local governments in France has been compared to Mauriac's Genetrix, who prevents his child from being a man while scolding him for not becoming one.[1] Local authorities are hardly ever involved in decisions about major national policies but are simply expected to support them. Nuclear policy is no exception.

The power of the highly centralized administration in France derives in part from the administrative and political fragmentation of French society. France is divided into about 38,000 communes. In one sense this allows a great deal of civic involvement; France has about 470,000 municipal councilors, or about 170 for every 10,000 voters (about five times as many proportionately as in the United States). But important decisions are made in Paris and implemented through the authorities of the 95 departments appointed by the central government and the departmental services of the technical ministries (such as equipment, industry, environment). This overwhelming power of the central bureaucracy over local government is justified by the jacobin and republican tradition which sees the central government as the guardian of the general will, transcending the politics of special interests.[2]

Absence of local responsibility has perpetuated resistance to social change, cynicism toward government, and nonpartisanship at the local level. While political concern is high, apathy *(qu'est ce qu'on peut faire)* is prevalent, reinforcing arguments for continued centralization, while encouraging resentment of state intervention.

In this context French nuclear policy evolved from consultation among the few ministerial departments concerned with industrial siting and regional planning: the ministry of industry, EDF, CEA, the ministry of equipment, the ministry of finance, and the ministry of planning. At first EDF was able to choose potential sites with few constraints. The departmental services of EDF and the technical ministries carried out preparatory work, and the prefect, as a buffer between rival claims of local interests and national policies, negotiated privately with local authorities to assure their compliance. These negotiations, usually involving commitments to certain material advantages for those communities willing to accept a nuclear power plant, were crucial for later public acceptance. Most mayors are receptive; although locally elected, they often derive their legitimacy less from their local constituency than from their ability to provide such material advantages via their relationships with the central authorities.[3]

France is the only OECD country that has not developed an atomic energy law to govern licensing procedures. Safety studies and operational control are exercised by departments within EDF and CEA. But increasing criticism of nuclear policy and the growing number of power plant siting disputes brought new actors into the administrative-planning process. DATAR, created to coordinate ministerial action in order to reduce economic imbalances in regional planning, and the ministry of environment, formed in 1971 to control environmental impacts, called for environmental studies and more detailed consultation with regional assemblies and local officials. Later, after the oil crisis, the government also created special agencies for energy conservation and for exploiting new energy sources, and the ministry of industry developed its own research capacities for energy planning. As the administrative structure governing nuclear decisions became more complex, so did its interaction with interest groups. Industrial lobbies concentrated on the ministry of industry and EDF; ecology groups directed their efforts to the new ministry of the environment. But also this ministry uses ecology groups to increase its bargaining power in intra-administrative conflicts with the ministry of industry or with EDF. In these often Byzantine and increasingly adversarial relationships, it has become more and more difficult to maintain the traditional image of a neutral administration, pursuing and defending the general interest against the objectives of particular groups.

The German Administrative Style of Consensus and Neutrality

In Germany implementation of national policies in most areas requires political accommodation with the elected *Länder* and local authorities. The present federal structure derives from the nineteenth century when the unification of the nation brought about by Bismarck left administrative autonomy to the *Länder*. The Weimar Republic reinforced centralized authority, but the postwar reconstruction of political life, influenced by the Allied occupation and the Anglo-Saxon ideology of local democracy, strengthened the federal tradition and the pattern of administrative decentralization.

The ten *Länder* of the Federal Republic vary in size and population, from Bremen with 700,000 inhabitants to North Rhine-Westphalia with over 17 million inhabitants. Within each *Land* local government units in the communes provide many services, and local authorities have associations with representatives on federal advisory councils.

At first national nuclear policy was the exclusive responsibility of the federal atom ministry. The first atom law of 1959 stipulated that nuclear plants are subject to specific licensing procedures that were later complemented by radiation protection regulations *(Strahlenschutzverordnung)* in 1960 and 1964. A commission to implement the regulations, chaired by the president of the German labor unions (DGB), Ludwig Rosenberg, was located in the atomic ministry. Eventually in the 1970s the regulation and safety of nuclear power shifted to the ministry of the interior which has overall responsibility for the environment.

The transition from research and development to commercial use of nuclear power led to more complex arrangements. The ministry of economic affairs sets the framework for energy policy in collaboration with the utilities and the *Länder*. The ministry for science and technology is responsible for R & D and for experimental plants such as the fast-breeder project in Kalkar. The ministry of the interior houses two committees: the nuclear safety committee, Reaktor-Sicherheitskommission (RSK), and the radiation protection committee, Strahlenschutzkommission (SSK). They supervise the *Länder* authorities who license power plants and the utilities that run them.

This formal organization, as well as the highly instrumental attitude toward problem solving in Germany, has enhanced the power of bureaucratic authority at the expense of parliamentary control. The administration plays an independent role in preparing and implementing economic, social, and regional development policies. This is especially true in nuclear policy where administrative services prepare energy forecasts, plan the nuclear share of primary energy sources, decide on licensing applications, and monitor compliance with emission regulations. In crucial decisions the services responsible for promotion and regulation work together. The perception of bureaucracy as the neutral custodian of the general interest against the particular claims of individual citizens and interest groups masks the political-bargaining process that is embodied in such techno-administrative decisions.

The German administration has traditionally sought to preserve its neutral image through the concept of social partnership in which the state integrates representative employers' and employees' organizations in making economic and social policy. This concept assumes the existence of organized groups with well-defined class interests. However, in environmental and technological areas such groups have not emerged, and social partnership has not worked to create political harmony. Indeed, as the bureaucracy grew, different agencies put forth contradictory expert opinions and forecasts concerning the need and the safety of nuclear power.[4] As in France the different federal agencies have mobilized their clients to support their agency's administrative objectives.

The consensus-oriented German administration differs in important ways from the French elitist and technocratic tradition. Yet, as these administrations change to deal with advanced technologies, they begin to converge. In both countries the increase in regulation has created internal conflicts that are especially sharp in the field of nuclear energy, where government is responsible for both promotion and control. The bureaucracies responsible for each task must develop a clientele to increase their intra-administrative-bargaining power. The need for such political support is expressed in both countries through administrative reforms—public inquiries and consultation procedures—intended to expand direct public involvement and win public acceptance of nuclear siting decisions.

Administrative Reforms

Technologies are reconciled with the values of a society through administrative procedures. Both the French and German governments have sought to expand and improve the channels through which affected groups could express their concerns about nuclear plant siting decisions.

The French government seeks public approval through regional consultation and public inquiry procedures. Regional assemblies were established in France in 1972 to serve as mediating bodies between the communes and the central government. In 1975 EDF submitted a list of thirty-four potential nuclear sites to these assemblies for evaluation and approval. This list included a dossier of technical information about the geological characteristics of each site, the energy needs of the region, and the safety aspects of the technology. The assemblies were to comment only on the acceptability of the specific proposed sites, not on the general acceptability of nuclear policy. Most gave their enthusiastic support: "Recourse to nuclear energy is absolutely necessary" (Aquitaine); "It is essential to the economy" (Poitou-Charentes). However, those dominated by the left opposition parties (Languedoc-Roussillon, Provence-Côte d'Azur, Midi-Pyrénées, Nord Pas-de-Calais) were concerned about general policy questions and asked for a broad public debate. One region (Centre) asked for expansion of the existing sites on the Loire and proposed a new site (Belleville) not appearing on EDF's list. Two regions (Alsace, Bourgogne) did not respond at all. Critics called this consultation effort a phantom exercise.[5] The assemblies had to use the technical dossiers provided by EDF to evaluate the sites: lacking the expertise and budgetary autonomy to organize an independent investigation, they could only respond according to existing political preconceptions.

At the local level the government decided that a public inquiry procedure normally required only for land expropriation should be extended to all nuclear plant siting decisions. Nineteenth-century legislation regulating conflict between private and public ownership required a Declaration of Public Utility (DUP) to expropriate land. The DUP involved a public inquiry. Prior to 1976 EDF and other governmental administrations had to use this procedure only in cases where they did not already own the necessary land and where the land taking was contested. Although EDF did in fact hold public inquiries, this was not required by law until 1976 when legislation required DUP inquiries at all nuclear sites. The dossier for such an inquiry must include an environmental impact study and a justification of the need for the proposed plant.

A construction permit also requires an Autorisation de Création (AC), approving the safety aspects of the nuclear plant, and special permits to assure the protection of historic monuments, establish compatibility with the interests of national defense, and coordinate construction with the air corridor pattern. An operating license requires a second series of procedures to assess environmental impacts.

Responsibility for the DUP public inquiry rests with the prefecture. EDF deposits its technical reports in the prefectorial offices and city halls of the concerned municipalities. People living within 5 km from the site have access to these documents for six to eight weeks, during which time they can voice their objections. At their early inquiries, before 1975, EDF had received only a few hundred objections. But the DUP inquiry at Blayet in December 1975 generated 23,000 objections. Also before 1975, when the DUP was perceived as a means to defend private property, objections had centered on the rights of ownership and the decline in property values associated with nuclear power plant construction. Subsequently objections have focused on pollution, safety, and other social and environmental concerns.[6]

The DUP is organized by an investigating commissioner, usually a local dignitary or a retired civil servant appointed by the prefect. He must be available to receive public objections for three days. Collecting the objections and evaluating EDF's response, the commissioner's report is the basis of the final government decision. No commissioner has ever denied an EDF application.

The DUP procedures have failed to build support for EDF decisions.[7] Critics find the technical data in the dossiers incomplete, lacking adequate information on questions of pollution and nuclear safety. They contend that political pressures prevent official criticism. For example, the Service de Sécurité des Installations Nucléaires (SCSIN), a division of CEA, was initially critical of the plan to site a plant in Le Pellerin because of its proximity to a city but later under political pressure withdrew its objections. Critics debunk the arbitrary choice of the inquiry commissioners, who often lack the technical competence to evaluate dossiers. They feel that public access to documents is restricted by offices open only at working hours and by the confidentiality of EDF's response. They are especially critical when EDF starts preparatory work on nuclear sites before the end of the inquiry procedure. This is within the law: evacuating and paving a

site can begin without a construction permit. But to critics costly preparatory work creates pressure on regulatory administrations to approve EDF
plans and indicates that EDF as a public enterprise does not expect to be
seriously challenged.

Criticism is reinforced by cases where sites are approved in the face of
obvious local resistance. In 1977 a public inquiry opened at Le Pellerin
after a long battle to prevent preparatory site surveys. The mayors of seven
out of twelve communes refused to use their offices for the inquiry, and in
two city halls documents were stolen and burned. The prefect was forced
to open an annex to the city hall on EDF property under police protection.[8] Eighty to 95 percent of the population in each of several municipalities signed petitions (one launched by the labor unions had 30,000
signatures) and agreed to boycott the official inquiry. Of the few who did
participate in the inquiry, 95 approved the project, and 750 opposed it.
During the incidents the prefect sued several peasants for destroying official documents, and the local court judged five of them guilty. But after a
spontaneous display of community support the peasants were released by
a higher court. The outcome of the inquiry was as unexpected as the whole
procedure: the inquiry commissioners, after declaring themselves incompetent to judge the issue, concluded in favor of the power plant.

This is an extreme example; even officials in EDF agreed that a caricature of participation is more counterproductive than no participation at
all. But while ecologists debunk inquiries as a democratic cover for policies
decided without public control, EDF contends

We have provided more information to the public than we are obliged to
do. We have innovated in the field of public participation, but everybody
criticizes us.[9]

EDF fears that the increase in procedural complexity will open opportunities for legal obstruction. It is especially concerned about the legislative requirement, effective in January 1978, to add an exhaustive environmental impact statement to the inquiry documents. This must include an
analysis not only of environmental impacts and the provisions to minimize
them but also a justification for the choice of a particular site as compared
to alternative possibilities. This has been especially controversial, for some
environmentalists have argued that, to justify the choice of one site, EDF
must investigate alternative sites as well.

Prior to this environmental legislation the DUP procedure was not
viewed as an adversarial process in the sense of a U.S. nuclear regulatory
hearing. There was no cross-examination nor provision for appeal; the

DUP was conceived more as a channel of information than a means to balance decision-making power. The system was designed to avoid political interference and allow the EDF to implement its policies insulated from the political process. The new environmental legislation, modeled on the U.S. National Environmental Protection Act, provides technical guidance as to how a policy should be implemented and allows for judicial and public review of administrative action. Thus it threatens to break the tradition of relatively closed administrative procedures.

In Germany the atom law (as amended in 1976) defining the procedures for licensing power plants also requires a public hearing. In Germany's decentralized administrative context the *Land* administrations provide construction licenses. They are supervised by the two federal nuclear regulatory commissions (RSK, SSK) and assisted by experts provided by a technical supervisory organization, Technischer Überwachungsverein (TÜV), and the Institute for the Safety of Nuclear Power Plants (Institut für Reaktorsicherheit) in Cologne. Project proposals, the dates of the public hearing, and the location of documents must be published in the official journal of the *Land* and in the newspapers.

Documents are available for public inspection for a month in the offices of the licensing authorities and in a designated location near the proposed site. Anyone potentially affected by a project is entitled to object. Standing to raise objections does not depend on geographic proximity as in France, and the courts have accepted claims from people within 100 km from a plant site.[10] Those who have introduced written objections are entitled to attend the hearings. But the hearings are restricted to discussion of the nuclear aspects of the plant; other issues, environmental, economic, and social concerns, are considered irrelevant.

The final decision must be justified in an extensive analysis of the proceedings circulated to the applicant and all those who have introduced objections. After this the decision goes into effect in fourteen days. Separate licensing procedures assess the impact of a power plant on land use and on water and air pollution. While emission standards in Germany tend to be stricter than in France, there is no single procedure to coordinate the evaluation of environmental impacts. Each is assessed separately, allowing partial construction licenses to be granted before the completion of all impact statements.

In principle these procedures allow the integration of divergent opinions into the decision-making process, but as in France environmental activists are critical on a number of grounds.[11]

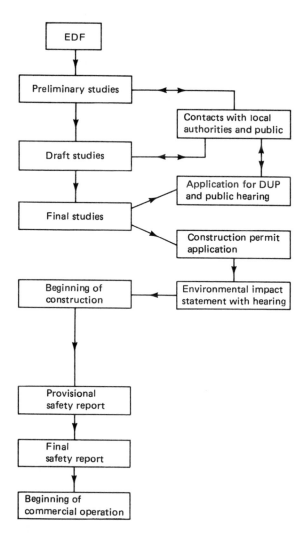

Figure 3.1
French siting process and licensing procedures (Source: Adapted from Michael Golay et. al., *Comparative Analysis of U.S. and French Nuclear Power Plant Siting and Construction Regulatory Policies,* Energy Laboratory report **MIT-EL 77-044-WP,** December 1977)

First they suspect the civil servants in charge of the organization of public hearings of bias, for they often serve simultaneously on the administrative boards of the applying firms. The neutrality of various supervisory and consulting bodies is also suspect. The ministry of the interior, responsible for supervising the *Länder* authorities, seldom questions their decisions. The research and regulatory laboratories consulted in the licensing process are tied to government and the nuclear industry. The large, state-run Institute for Reactor Safety was developed by the first generation of enthusiastic German nuclear experts. Its president is a member of the board of directors of the Hamburg electricity firm (Hamburger Elektrizitätswerke). The TÜV, created by industry in the nineteenth century to check the safety of steam vessels, now has a monopoly over the regulation of safety in very diverse areas. It remains a private association, but with recognition as an official regulatory body. Critics argue that despite its public function this private association has strong links with industry, for it is a member of the Atomforum, the lobby in favor of nuclear energy.

Second, critics question the adequacy of the procedures. Only those documents that are part of the formal application are available to the public. Internal administrative evaluations are open only on the discretion of the administration. Moreover as in France documents are only available during working hours and often written in language difficult for nonspecialists to understand.

Critics also contend that the multiple-licensing procedure obscures the relations between different environmental impacts and that the granting of partial licenses creates powerful economic pressure for approval. Finally, a number of incidents suggested that the procedures permitted little real choice. Confronted with demonstrations at Wyhl, the prime minister of Baden-Württemberg declared before the licensing procedures: "There can be no doubt that Wyhl will be constructed." [12] The *Land* minister of economics, politically responsible for the procedure in Wyhl, was also the acting vice-chairman of the utility's board of directors. In the case of a plant in Esensham (Lower Saxony) regional officials acknowledged that secret negotiations with the nuclear industry were already underway for more than a year before the official application for a construction license.

In both France and Germany avoiding costly delay is a central priority in the licensing procedure. Civil servants, convinced that their plans serve the general interest, prefer closed procedures involving participants from industry and government who are accustomed to mutual accommodation. Integrating the public into this decision-making process means only in-

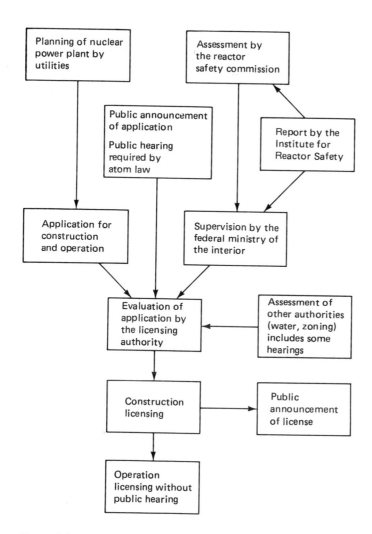

Figure 3.2
Nuclear power plant licensing procedure in West Germany (Source:
Bundesministerium für Forschung und Technologie, *Viertes Atomprogramm der
Bundesrepublik Deutschland für die Jahre 1973–76*, Bonn, p. 80)

creased delays. Thus inquiries and other participation channels are considered more an undesirable necessity than a serious responsibility. "Getting through it" is more important than dealing in depth with public objections. They support administrative reforms, but mainly to provide legitimacy. Thus they are directed more toward co-opting public support than to changing the decision-making process—more toward seeking informed consent than expanding democratic choice. Alain Peyrefitte, once minister for administrative reform, quotes the departmental director of a technical service to illustrate administrative attitudes toward public consultation *(concertation)*:

We are obliged to make decisions. The elected officials know nothing. They aren't capable of discussing our projects with us. We don't talk the same language. The concertation is very nice; reality is something else. In the end we decide what to do but give the officials the impression that it is they who decide. One must play the game.[13]

Determination to implement preconceived decisions leads officials to ignore, to debunk, or simply to be unaware of political opposition. But it leads critics to view participation procedures as rhetoric in a closed system where decisions, sheltered behind the image of a neutral administration, continue to rest on collaboration between the bureaucracy and industry. Their impressions have been reinforced by the very limited role of traditional political institutions in the decisions concerning nuclear power.

4 The Problem of Political Control

The decisions to expand nuclear power in France and Germany were of major social and political significance. Yet nuclear policy hardly engaged traditional political institutions at all. Only when the strident character of the controversy reached a high pitch in 1977 did the political parties and the unions become significantly engaged. Then the large-scale demonstrations and the ecologists' electoral success finally evoked a cautious response. The passive role of the representative institutions is a striking and important aspect of the nuclear debate, establishing the conditions for the extraparliamentary character of the dissent.

Parliamentary Impotence

Public policy in France is formulated by the president of the republic and the government; parliament serves less to initiate legislation than to supervise policy. During the period of rapid governmental changes in the 1940s and 1950s administrative bureaucracy, not parliament, appeared as the stabilizing factor in French political life. Subsequently its power has increased, so that parliament is not likely to challenge administrative decisions. The membership of parliament has reinforced this relationship. Between the Fourth and Fifth Republics the proportion of former state bureaucrats tripled, until half of the ministers and 90 percent of their staff were former civil servants.[1] Parliamentary power is limited by its relatively poor technical facilities; it has few specialized commissions or advisory groups and little staff assistance as compared to the U.S. Congress. The well-established presidential majority and its unchallenged tenure since 1958 has also precluded meaningful parliamentary involvement in controversial policy areas such as nuclear power. In the absence of special legislation controlling this technology, parliament has had no important role.

The government introduced the major decision to expand its nuclear program in the fall of 1974 as a fait accompli. Parliament did not vote on this issue, which was defined by the government as a technical implementation of energy policy, needing no special debate. Although most parliamentarians accepted the nuclear choice as an economic necessity, many of them expressed their resentment of this closed decision-making power. At a parliamentary debate called to discuss the government decision, a socialist representative criticized the regional consultations over nuclear plant siting as hypocritical pseudoconsultation and charged that they confused information and propaganda and stifled debate. Consultation procedures, he argued, reflected a scorn for democracy. One Union pour la Démocratie Française (UDF) representative favored the expanded nuclear program as a logical extension of the Fifth Republic plan to develop economic independence and dismissed fear as an irrational leftover from Hiroshima. But another proposed instituting a supreme court for safety to assure that information was not simply propaganda and that safety measures were enforced. Wide agreement prevailed on the need for more extended discussions with elected officials, regional leaders, and environmental associations.

Political sensitivity focused less on the substance of nuclear policy than on parliamentary impotence concerning this emerging political issue. Only in 1977 did parliament begin to develop a more substantive critique. The parliamentary commission for finance criticized the economic implications of government energy policy, arguing that the plan for an all-nuclear economy could maneuver France into a new situation of energy dependence.[2]

The German parliament is also limited in its influence over public policies. Major decisions are made in party headquarters through complex consultation procedures intended to establish a consensus among representatives from the *Länder* and the key interest groups. The hope is to avoid conflict within the establishment; thus by the time an issue gets to parliament, support is often unanimous.

Parliamentary power is limited in Germany for different reasons than in France: the German parliament can create commissions and organize hearings, and legislators have better access to technical advice than their French colleagues. The Germans vote for parties rather than for individual representatives, and only the parties can present candidates for election. This procedure imposes serious party constraints on the behavior of parliamentarians. Even where the consensus is fragile, as in the case of nuclear policy, the party oligarchy controls its constituency. It is politically risky to express opinions that oppose the consensus as dissenters are expected to display unanimity or to resign.[3] For example, a CDU (Christian Democrat) spokesman for environmental problems left the party in 1978 because of his antinuclear position. He formed a separate conservative ecology party. In another case six liberal members of parliament who refused to vote for continuing construction of the Kalkar fast-breeder reactor were forced to return to party discipline. As in France the presence of a high percentage of former civil servants in the legislature (in 1979 some 50 percent of all legislators in the national and *Länder* governments were civil servants) reinforces parliament's tendency to accept administrative proposals.[4]

However, unlike in France the German parliament has actively debated nuclear policy, and in 1960 it adopted the atom law. But parliamentary attitudes toward nuclear power during the early years of the nuclear program were based on what Willy Brandt described as a "blind belief that the nuclear problem was solved by renouncing atomic weapons and agreeing on the atomic weapons control treaty." [5] The Social Democratic party's (SPD) first choice of an official spokesman for nuclear policy reflected this cavalier attitude: he was the member of parliament from Karlsruhe, the location of the most important nuclear research center—a man neither familiar with nuclear power nor anxious to be nominated.[6] The issue of potential conflict of interest was not considered, for nuclear power was not yet viewed as a problematic political issue.

In its deliberations parliament has relied on the advice of experts from public research centers and the nuclear industry. Substantive discussions have focused on traditional political and economic questions—the implications of government control over industrial practices, the potential military use of a developed nuclear capacity, and the domestic economic consequences of international agreements. But throughout the 1970s most discussions were of little interest to legislators, who felt out of their depth when asked to debate an issue presented as a technical option, and few were involved. For example, in July 1975 the minister of the interior presented a document to parliament explaining nuclear policy. When it was finally discussed at a late night session at the end of January 1976, less than 50 of the 518 parliamentarians were present.[7]

When parliament did debate the issue, it faced unusually heavy industrial lobbying, a problem in the German political context where lobbying is negatively regarded as violating the autonomy of the legislator. During a proposed reformulation of the atom law in 1976 the nuclear industry lobbied against a bill to introduce liability, and electricity supply firms lobbied for a bill to license general plant models so as to avoid individual-licensing procedures at every site. These pressures failed to change the legislation because of intense counterpressure from the citizen initiative movement.

From 1972 to 1976 several legislators had active environmental interests. The minister of the interior, Hans Dietrich Genscher, responsible for
environmental protection, tried to present the Liberal party (FDP) as an
environmental party. The environmental spokesman of the opposition
party, CDU, Herbert Gruhl, the author of a best-selling book on the limits
to growth, had a personal reputation that allowed him some independence
within a party representing mainly industrial interests.[8] Gruhl left the
CDU in 1978 to create the ecological party Green Action Future (Grüne
Aktion Zukunft). Similarly Frank Haenschke, the technology spokesman
of the SPD, despaired of parliament's ability to influence technology policy and decided not to stand for re-election in 1976.

Haenschke's judgments were severe: parliament is nothing more than a
working committee, and its discussions are but a ritual *mise en scène*.
Actual decisions are reached in committees where experts reconcile their
differences outside of public view. While this is efficient, the outcome of
committee discussions, he claims, depends less on political judgment than
on expertise; what matters is the strength and size of the specialized armies
that protagonists can mobilize.[9]

The Wavering Party Line

Political parties act as intermediaries between the national government
and the citizenry. One would thus expect them to serve as channels for
organized conflict resolution in the intensely politicized area of nuclear
power. Yet in both France and Germany they played only a marginal role
in the nuclear debate.

In France, with its fragmented political culture, interests are generated
and articulated more within governmental structures than in the parties.
Only the Gaullists and Communists maintain a coherent, pragmatic argument about nuclear policy; and these are also the only French parties with
continuity in organization, leadership, and membership.

In the centripetal political culture of Germany the need to create consensus among elites limits open-party debate on important issues.[10] The
different elements of the political system are interrelated through multiple
membership and overlapping leadership in the parties and major interest
organizations (labor unions, professional groups, and employers' and
farmers' organizations). If the preferences of these organizations coincide
with administrative policy, as in the case of nuclear policy, the political
parties normally endorse their decisions, seldom articulating interests that

run against the will of the major groups. Because of these structural relationships the political parties in both countries avoided the issue of nuclear
power until dramatic events in 1977 forced them to respond. Even then
their internal divisions created such considerable ambivalence that they
proceeded with extreme caution.

The activities of the political parties in France reflect the complexity of
her multiparty system and the tendency since the early 1970s to form coalitions. The Gaullists and the communists have each enjoyed the solid support of about 20 percent of the electorate. The Union pour la Démocratie
Française (UDF) is a coalition among several small center parties that collaborate with the Gaullists in the presidential majority. The Socialist
party, consolidating several heterogeneous groups, formed an electoral
coalition with the Communist party in 1972.

These parliamentary coalitions did not entirely abolish programmatic
cleavages: the partners in the left union and in the presidential majority
have diverging views on such crucial issues as European integration,
NATO, and nuclear weapons. These schisms assumed importance as the
parties tried to establish a consensus on nuclear policy.

The Gaullist party (RPR) is the only major party to unequivocally ad-
vocate expansion of nuclear power in France.[11] Once a promoter of military nuclear policy, the RPR sees the rapid expansion of the nuclear program as a means to assure energy independence and the participation of
French industry in international competition. Favoring a strong centralized state, the Gaullist party also views the decentralizing ideology of
antinuclear groups as a threat to national unity.

The other parties view nuclear policy with ambivalence. The UDF
reflects the rivalries of its three right-center parties, the Centre de
Démocrates Sociaux (CDS), the Radicals, and the Parti Republicain (PR),
each with different views on nuclear policy. CDS is critical of nuclear
power, favoring the implementation of a minimum program and the development of new alternative energy sources.[12] Some of its prominent
members, such as Philippe St. Marc, are environmental advocates of technologies allowing local-level participation.[13]

The theme of decentralization is also attractive to both the Radicals and
the PR, but they are pro-nuclear. The Radicals criticize the secrecy and
centralization of decision making characteristic in the nuclear sector, but
they also criticize voluntary associations for obstructing the public interest. The PR, a typical party, or rather nonparty, of French conservatism,

uncritically supports nuclear energy and the interests of the nuclear industry.[14] Shaped by these ideological differences among its constituent groups, the UDF is necessarily ambivalent toward nuclear policy.

The Communist party (PCF) has generally supported nuclear technology but has at times criticized its implementation. Defending the concept of state-controlled R & D, the party perceived the government's 1969 decision to drop the French gas-graphite line in favor of the American light-water reactor as a policy to abandon the nationalized sector and subsidize private French and multinational corporations. For the PCF this development demonstrated the dependence of France on U.S. and German imperialism and documented the policy control exercised by oil companies and the nuclear industry:

Giscard d'Estaing . . . justifies the policy of "all atomic" for the profit of Baron Empain and Westinghouse in the name of national needs and national independence . . . . The safety of the population counts no more here than at Seveso or at Minimata.[15]

The Communist party also criticized the lack of democratic control over decisions, the absence of parliamentary debate and the truncated dossiers given to local officials.

Despite such reservations the PCF sees this technology as a means of promoting better conditions for workers and achieving the economic and energy independence necessary to accomplish the transition to socialism.

The Socialist party (PS), heir of the former social democratic SFIO (Section Française de l'Internationale Ouvrière), was created in 1969 with a heterogeneous social composition. Its long-term objective is to create a self-determination model of socialism. This ideology, called *autogestion,* is rooted in both a radical libertarian social analysis and the Christian socialist tradition. The continued expansion of the progressive wing of the Catholic church over the last fifteen years, as well as the influx of younger activists, has reinforced this concept of decentralization.

In 1976 the party established an internal expert energy commission, led by several EDF officials. The commission first argued in favor of a nuclear program, but some of its members urged the party to remain open to the growing environmental movement. The party wavered, but the antinuclear arguments were taken more seriously after the demonstration in Creys-Malville. By then competition from ecological candidates during the municipal elections of March 1977 converged with pressure from members active in the Confédération Française Démocratique du Travail (CFDT), forcing the PS to make a decision. A working group of economists, sociologists, and engineers failed to reach programmatic agreement. The majority report argued for a moratorium and a policy of phasing out the nuclear option, but a minority insisted on France's need for nuclear energy.[16] Forced to act during the pre-electoral period of fall 1977, Socialist party leader François Mitterand announced that the party favored a two-year moratorium on nuclear development. After the 1978 elections the internal debate was renewed, mainly between those favoring high economic growth and increased energy consumption and those arguing for a slowdown of nuclear expansion for reasons of safety. The debate did not include the option of abandoning nuclear power.

The Socialist party maintains a parliamentary alliance with the Mouvement des Radicaux de Gauche (MRG). This splinter group from the radical center party is sympathetic to environmental groups and favors a nuclear moratorium. Because of its middle-class constituency, including a large number of local officials from the center and the south of France, this party has more influence than its limited parliamentary representation would suggest. Similarly the Parti Socialiste Unifié (PSU), on the left of the Socialist party, has taken a definite ideological position against nuclear power.

The radical left parties were the only ones to actively participate in antinuclear activities. Maoist and Trotskyist groups have supported many activities of the ecology movement, warning that nuclear energy implies a repressive police state and control by multinational firms. These parties occupy a marginal, but not negligible, position in the French political spectrum (with a 2 to 4 percent share in national elections). While having no direct political influence, their youthful constituency often captures the attention of the major parties.

On the surface the German political parties offer a simpler picture: the two main German parties, the Social Democrats (SPD) and the conservative Christian Democrats (CDU), with its Bavarian branch, the Christian Social Union (CSU), are remarkable for their consistent strength. For over two decades these parties have each received at least 38 to 42 percent of the votes. Only the Liberal party (FDP), comprising from 5 to 10 percent of the electorate, has survived outside of this two-party system. Since the 1950s programmatic and ideological differences between the two major parties have progressively diminished as both seek to be mass organizations (Volkspartei) representing a span of social class interests.

Despite this homogeneous image differences over economic and international issues persist. These were reflected in the early debates over nuclear policy. The SPD opted for state control of energy development when CDU/CSU gave priority to private industry; the SPD wanted Germany to renounce nuclear military development when the CDU/CSU tried to keep this option open; the SPD urged signing the nonproliferation treaties when CDU/CSU wanted changes allowing greater national control.[17] Major ideological differences also persist within the parties, making it difficult for them to come to terms with the nuclear debate.

Among the conservatives the Bavarian CSU is the strongest supporter of the nuclear industry. Lacking coal resources, Bavaria has faced serious energy problems. The first atom ministers, Franz Josef Strauss and Siegfried Balke, came from this region. However, the CSU, in a conservative tradition that places a high value on nature, also favors environmental protection.

The CDU, with a relatively strong labor wing *Sozialausschüsse,* favors nuclear power, and the harshest police reaction to antinuclear demonstrations has come from CDU governments in Schleswig-Holstein, Lower Saxony, and Baden-Württemberg. CDU blames such unrest on the reform euphoria and unrealistic expectations created by the social-liberal coalition. The party is also ambivalent about citizen initiative groups, explaining them as a reflection of parliamentary weaknesses, while criticizing them for their negative economic effect and obstruction of technological development.[18] It is most critical of the nationally organized Bundesverband Bürgerinitiativen Umweltschutz (BBU), which has become the leading antinuclear organization.

The SPD has also faced internal conflict in developing an energy policy. Many of its local and regional organizations, for example, in Hamburg and Baden-Württemberg, oppose nuclear power siting proposals; but labor is an important SPD constituency, and the major unions endorse nuclear policy. Cleavages were revealed at a major conference, Energy, Employment, and the Quality of Life, in April 1977. The entire party leadership, including Willy Brandt and Chancellor Helmut Schmidt, attended, suggesting the importance of this issue. The key discussion paper by Erhard Eppler (former minister for development aid and head of SPD in Baden-Württemberg) strongly questioned whether in a democratic system one can ignore the significant public movement against nuclear power.[19] The trade union representatives and several ministers, Hans Matthöfer (then minister of research and technology) and Herbert Ehrenberg (minister of work and social welfare) favored the nuclear program. However, the party youth organization Jungsozialisten (JUSO) opposed nuclear power, attacking the monopolistic character of the industry.[20]

After this conference the SPD continued the discussion in its local organizations and newspapers; the predominant attitude expressed in this dialogue indicated that SPD would vote for a two-year moratorium at the party congress in November 1977. Then one month before the congress 40,000 trade union members, mainly from the nuclear industry (KWU) and the firms engaged in plant construction (Krupp, Hochtief, Mannesmann), participated in a pronuclear demonstration in Dortmund.[21] Orchestrated by a public relations agency from Munich, and very well publicized, it forced the SPD into a compromise. At the party congress SPD accepted the trade union position but stressed that a solution for waste disposal must be found before further nuclear expansion.

The Liberal party (FDP) at first proposed a moratorium on the nuclear program, but at its congress in November 1977 a small majority voted for the government program, favoring further nuclear expansion once the waste disposal problem is solved. The split in this party went along traditional lines: the conservative wing sought to protect industrial interests; the youth organizations argued that civil rights would be endangered by a plutonium economy.

Despite the cleavages the debates at the 1977 party congresses were far less dramatic than expected. The leaders of FDP and SPD urged their members not to threaten the fragile majority of the government coalition in parliament through antinuclear activities. Committed to an image of consensus, the internal ideological differences never even reached the political arena. Moreover the terrorist activities in 1977 also diverted the discussion, as the nuclear issue took second place to the theme of internal security. In the 1979 party congresses the government coalition parties reopened the issue in preparation for the 1980 electoral campaign. Their decisions basically reflected their continued ambivalence toward nuclear power.

By late 1977 dramatic violence at antinuclear demonstrations in both France and Germany, and the success of ecology candidates in the election campaign in France, forced serious discussions of the nuclear issue and revealed the ideological ambiguities and organizational constraints that had kept the parties out of this major political debate. The questions raised by the nuclear debate cut across party lines but were internally divisive as well. Participation in the nuclear debate could only exacerbate internal differences as attitudes toward nuclear policy were split (see table 4.1). Internal debate was especially divisive in the SPD and FDP where party leadership confronted an antinuclear majority within the party membership.

In Germany SPD's ambivalence about nuclear power partly reflects its traditional attitudes toward technological innovation. Closely tied to the labor unions, SPD has favored technological change to improve the conditions of the working population. But at the same time technology poses problems of labor substitution and occupational safety that temper its generally positive view.

SPD's ambivalence has been reinforced by changes in the party's social composition: a managerial elite joined the SPD after its transformation into a *Volkspartei,* hoping to modernize German economy and society and free it from the stalemate of the CDU state.[22] But the SPD also inherited activists from the 1960s student protest movement who stressed public control of the economic apparatus and ecological themes. These new con-

Table 4.1
Party preferences and attitudes toward nuclear energy in Germany and France

Party preference	Con	Pro	No answer or undecided
Germany[a]			
SPD	45%	38%	17%
FDP	52%	39%	9%
CDU/CSU	37%	38%	25%
Other	40%	31%	29%
Total population	41%	37%	22%
France[b]			
Left			
Communist party	57%	33%	10%
Socialist party	48%	46%	6%
Majority			
UDF (Giscardists)	26%	68%	6%
RPR (Gaullists)	26%	68%	6%
Other	50%	30%	20%
Total population	42%	47%	11%

[a] Source: INFAS 1977. In this survey people were asked if the safety of the population was endangered by nuclear power plants.
[b] Source: *Figaro* SOFRES 1978. In this survey people were asked to indicate their position toward nuclear energy on a six-point scale, ranging from "strongly against" to "strongly in favor."

stituents account for cleavage within the SPD, as former student activists confront the modern managerial elite. Their diverging interests nourish the tension that pervades discussions of energy matters.

Similar diversity confronts the left in France. The Communist party, with an estimated 300,000 to 500,000 members, is still mainly working class, but many professionals and intellectuals attracted by the party's less dogmatic policies joined after 1968.[23] The Socialist party, with about 150,000 members, includes such ideologically diverse elements as the old social democratic activists, the liberal intellects who joined the republican clubs in the 1960s to elaborate an alternative to Gaullism,[24] young people from the May 1968 experience, and activists from the CFDT. It especially appeals to professionals, technicians, and civil servants in middle- and high-management positions. This class is the most divided over nuclear power, for both ecological and technological goals are elements of a modern ideology.

In each country the left must cope with the divergent viewpoints of labor, the technocratic-managerial middle class, and the activists inherited from the student movement. In Germany cleavages appear inside the SPD, while in France the differences are mostly reflected in the fragmentation of the left.[25] But in both cases silence was convenient to avoid conflict.

Trade Union Ambivalence

For several decades European trade unions have developed positions on social and economic policy, and their views on major new technologies such as nuclear power are politically important. The extent of union membership varies widely in western Europe, ranging from 85 percent in Sweden to 23 percent in France. German unions include 42 percent of the workforce. France has the lowest level of labor organization in Europe.

The organizing principles underlying French and German trade unionism also differ. In France unions compete over ideological lines and political affiliations, whereas the main German trade unions are united in a single federation, Deutscher Gewerkschaftsbund (DGB), comprising more than 82 percent of organized labor. The German union structure is rooted in the experiences of the Weimar Republic when the trade union movement, fragmented into confessional and political groups, had been incapable of opposing the rise of national socialism. It has since been understood that only a strong and united federation can defend workers' interests and assure their participation in national policy.

European unions are closely related to their corresponding parties. The two most important French unions, the pro-communist Confédération Générale du Travail (CGT) and the Confédération Française Démocratique du Travail (CFDT), have close links with the left parties and support their goal of nationalizing key industrial sectors and the banking system. Both unions in the long run hope to transform France into a socialist country. But the CGT favors centralized solutions, regarding state control by labor organizations as the best guarantee for social justice, while the CFDT favors decentralized solutions based on the decision-making autonomy of basic social units *(autogestion)*. This difference has shaped their views of nuclear power.

In Germany the DGB has strong ties with the SPD, the socialists providing by far the greatest support in union elections. The DGB and the employers' association, Bundesverband der Deutschen Industrie (BDI), derived considerable legitimacy from their contribution to postwar reconstruction. Their spirit of compromise and determination to avoid social conflict are often invoked to explain the economic miracle of postwar recovery, and both BDI and DGB are consulted on a wide range of economic and social problems. Their representation in *Konzertierte Aktion*—a consultative body set up in the economic ministry at the end of the 1960s—institutionalized their influence on national economic policy.

The integration of the German labor movement into government policy making, however, does not necessarily reflect the political orientation of the most active unionists. The vote to nationalize key industries at the DGB general congress in 1978 demonstrated the persistence of Marxist traditions. This decision was withdrawn following a compromise among the most powerful union leaders, but the younger generation of activists have sustained their radical ideology.

Although many blue-collar workers are ambivalent about nuclear power (see chapter 8), the largest labor organizations in both France and Germany favor the technology. Their support of nuclear energy reflects their preoccupation with immediate economic issues such as unemployment. Thus their criticism of nuclear power is limited to details of the government programs and procedures and to the role of multinational corporations.

In France the CGT has criticized the secrecy of the decision-making process, the priority of profitability over workers' safety, and policies that favor multinational enterprises. However, CGT has assumed that further rapid economic growth and energy expansion is absolutely necessary to maintain a reasonable standard of living. "We can never relieve the problems and improve the existence of the millions of people now living in poverty without increasing their energy consumption." [26] In the communist tradition CGT regards ecology and zero growth as a new form of petit bourgeois "left extremist, an infantile disorder." [27] Thus until the Three Mile Island accident CGT unequivocally endorsed expansion of nuclear power. After the accident it demanded the strengthening of safety controls and a 50 percent reduction in the construction program, from seven to eight plants to only three plants a year.

CFDT takes a more explicitly critical position. Growing out of the Christian labor movement, and integrating Marxist analysis since the beginning of the 1960s, the CFDT is known for its unconventional and often radical positions. Its ideological openness encouraged the participation of activists of May 1968. Its concept of *autogestion* satisfies Christian socialist ideals of social harmony through small basic units and also appeals to radical revolutionary views in the anarcho-syndicalist tradition which has always been a strong minority element within the southern European trade union movement.

Other factors explain the different views of these two major unions toward nuclear power. The CGT, with its largely blue-collar constituency, emphasizes traditional pragmatic union demands: wages, purchasing power, national control over production. The CFDT, representing mainly white-collar workers, presents more qualitative demands concerning working conditions and occupational safety; it has been one of the few French political organizations to take up the themes of May 1968 and define them in practical social terms.

In 1975 a CFDT committee on nuclear energy offered the first extensive critical analysis of nuclear energy in France, a book that promptly became an important reference for the antinuclear movement.[28] Reviewing all the arguments that led the union to reject the 1974 Messmer plan, this book provides a technical explanation of nuclear technology and an economic analysis of French industrial policies. It describes the impact of nuclear power on working conditions and analyzes its implications for larger production units and for greater political and economic centralization. This extensive analysis does not conclude with a clear rejection of nuclear power but presents the governmental choice of *tout nucléaire* as a policy that works in the interest of multinational corporations at the cost of considerable public risk. After 1976 CFDT criticism, deepened by a long strike at the nuclear fuel reprocessing plant at La Hague (see chapter 6), became increasingly radical. It was the first large organization to support the ecologists' demand for a moratorium and eventually convinced the Socialist party to adopt this line. Then some of its activists participated in the 1978 legislative elections on the platform *Front Autogestionnaire,* formed mainly by Marxist ecologists and libertarian socialists of the PSU. After the Three Mile Island accident the CFDT demanded an immediate three-year moratorium. It also became the driving force behind the elaboration of an agreement, signed in June 1979 by a coalition of the PS, PSU, MRG, Les Amis de la Terre, GSIEN, and consumer organizations (the largest antinuclear coalition ever formed in France), asking for a large national debate on nuclear energy.

The diversity among the French unions contrasts with the apparently monolithic position of the German DGB. But the cleavages within the SPD also pervade the union. In the early days of German nuclear policy official statements from the DGB were cautious. Engaged in the campaign against the military use of nuclear energy during the 1950s, the DGB admitted the need to use nuclear power as a source of energy, but it stressed its dangers.[29] Once bomb testing had ceased in 1963, and military nuclear policy became a less prominent issue, only those sections of the union directly involved in the energy sector expressed any interest in nuclear power. After the dramatic demonstrations in Wyhl and Brokdorf, and just before the 1977 SPD conference on Energy, Employment, and the Quality of Life, DGB presented a pronuclear statement, but one that stressed the priority of safety over economic objectives and the importance of solving problems of waste disposal before further expansion of the nuclear program.[30]

DGB's caution reflected the different perspectives toward nuclear power among its constituent unions. These differences reflected their direct economic interests. The mining union argued for continuing the nuclear program in parallel with an increasing use of coal.[31] The construction union, motivated by employment objectives, opposed any reduction of coal or nuclear plant construction and avoided discussing the other problems involved in the nuclear debate. The chemicals unions, concerned with unemployment and the position of the German chemical industry on the international market, also backed the government program.

However, the metallurgy union, with the largest membership, has been critical. After the Brokdorf demonstrations its newspaper *Metall* published articles on the social and political implications of nuclear power. The newspaper also attacked the prevailing assumptions equating energy supply with economic growth and employment. Many letters to *Metall* supported its antinuclear position. But union representatives from factories directly engaged in supplying power plants (KWU, BBC) organized a pronuclear demonstration at the end of 1976 (with the slogan: "No power stations—No progress") and submitted a pronuclear petition signed by thousands of workers to Chancellor Schmidt. In the summer of 1977 they hired the same public relations company that had organized the mass demonstration at Dortmund before the SPD party congress to conduct a pronuclear advertising campaign to influence the labor movement.

The public services union has also experienced internal conflict. During the summer of 1977 union debates about problems of waste disposal and recycling revealed skeptical attitudes. But after the antinuclear demonstrations at Brokdorf, the workers from the utility, Nordwestdeutsche Kraftwerke, which was to build this nuclear power plant, organized demonstrations with the slogans "Safe energy for everyone," "It is a matter of our jobs."

Despite the ambivalence of major unions such as metallurgy and public services, the *coupe de théâtre* in the Dortmund stadium involving 40,000 unionists forced the DGB leadership to support nuclear policy. Nuclear critics promptly labeled this decision an *Atomfilz*—a word implying the collaboration of union officials with the nuclear mafia. H. Brandt, a popular union activist, defined this concept in a speech at the Brokdorf demonstrations:

By *Atomfilz* I mean those well-paid trade union bureaucrats on the supervisory boards and managing boards of the big companies who conspire with the nuclear energy lobby, and are repulsively entangled with them, to our shame.[32]

DGB officials who had reached their consensus were furious at such public criticism and tried to expel Brandt for his antiunion behavior. Brandt's rank and file support prevented such action, although one union leader who was both an SPD member of the parliament of Schleswig-Holstein and a member of the supervisory board of the Norddeutsche Kraftwerkunion sued him for insult, slander, and defamation. Popularity protected Brandt, but lesser-known activists are vulnerable. They complain that the union establishment tries to stigmatize nuclear critics by associating them with radical anticonstitutional forces and that unions even accept material delivered by the political police (the *Verfassungsschutz*), thereby extending to the unions the *Radikalenerlass* (laws preventing radicals from entering civil service jobs).

In France the unions, with their fragmented structure, are far less integrated in the political decision-making process than their German counterparts, and therefore their leaders have more freedom to express critical views. The main problem for the French labor movement is to find ways to move beyond their radical discourse and actually to exert influence on public policy. In contrast, the major problem for German unions is to maintain the tradition of union democracy without losing their existing ability to affect public policy.

This is a significant difference, and it determines the unions' relationships with new social movements and citizen initiatives. The French CFDT is open to such contacts, finding them useful in revitalizing attitudes toward the work environment. In Germany the oligarchical organization of the DGB reduces the possibility of collaborative action with citizen initiatives. Confronted with the growing importance of such groups, DGB officials regard their union with its 9 million members as "the largest German citizen initiative." This view ignores the active and growing participation of its younger members in the initiatives, and they find their interests poorly represented by union activities.

In sum important cleavages in French and German political institutions have greatly limited their representative role concerning nuclear policy. Issue-specific factors (the technical aspects of the problem, the monopoly of expertise in administrative bureaucracies, and the relationship between government and industry) further contributed to depoliticization of this policy area. Indeed the different political traditions and institutional structures characteristic of each country only serve to mediate the process through which this major political issue has been translated into a technical problem.

This depoliticization of the nuclear issue is part of a larger syndrome— the shift of politics from the level of parliament and the parties to that of administration and the bureaucracy. As we have seen in chapter 3, major problems requiring more than general political competence are increasingly handled by professional bureaucracies whose power is enhanced by the expanding field of government activity. This presents an important dilemma for western democracies, as the need of specialized expertise broadens the gap between the citizenry and the state and contributes to a crisis of legitimacy.[33]

As the traditional political organizations lose their representative function, public concerns are expressed less through conventional political channels than through new forms of organization and new modes of political action. Indeed the lack of involvement by the political organizations creates a fertile ground for new nationwide movements outside of the traditional political structure. This is what has happened in the nuclear power controversy.

II THE CYCLE OF OPPOSITION

5 The Escalation of Nuclear Protest

The antinuclear movement first developed more as a set of discrete and spontaneous events than as a coherent, consciously elaborated political campaign. But a flourishing ecology press widely publicized local actions, and parallel actions recurred in distant places. In retrospect we can distinguish several stages in the evolution of antinuclear protests in France and Germany, each characterized by a predominant type of political behavior.

The antinuclear movement began in the early 1970s with a series of local siting disputes that aroused public concern about the risks of nuclear power and established the right to resist. With the expansion of the nuclear programs in 1974 and 1975, political action turned from specific siting decisions to national nuclear policy, and the discourse turned from environmental preservation toward more apocalyptic themes. Concurrently the economic and political analyses deepened to emphasize the concentration of economic power stimulated by nuclear policy, the inequities of its impact, and the political problems in assuring adequate control.

Reflecting these changes, the second stage in the evolution of protest was marked by disruptive confrontation. Motivated by a moral fervor to protect justice, democracy, and indeed human life, the activists sought more effective actions to prevent the construction of nuclear power plants. "When justice is turned into injustice, to resist becomes an obligation!" This slogan of the German activists at Wyhl expressed the justification of the mass demonstrations and site occupations after 1976. The mood and the tactics during this second phase of the antinuclear movement left little space for negotiation, and polarization was inevitable.

The confrontations provoked an immediate response. The rules were clear: public authorities accept protest as long as it does not threaten crucial policy decisions. The occupation of proposed plant sites infringed on these limits. Obliged to protect the utilities, the government mobilized its police. The antinuclear activists had overestimated their own strength and underestimated the willingness of government to use force. The media response, initially positive toward the movement, now focused on violence and the infiltration of the movement by left radicals and professional troublemakers. The backlash undermined the efforts of environmentalists to mobilize public support and caused internal dissension within the movement. Thus after the dramatic and violent confrontations at Brokdorf and Grohnde in Germany and Creys-Malville in France, all in 1977, the movement turned to a third phase, exploiting more diverse strategies, including conventional political participation and nonviolent action.

A Right to Resist?

In 1970 a teacher, Jean-Jacques Rettig, created the first local antinuclear association in France to oppose a proposed plant in Fessenheim, Alsace (Comité de Sauvegarde de Fessenheim et de la Pleine du Rhin). The next year one of the first journalists for the ecology press, Pierre Fournier, and a teacher, Emile Premillieu, formed a committee to oppose the nuclear plant at Bugey-Cobayes.[1]

These were the sites of the first two antinuclear demonstrations in France. In April 1970 more than 1,500 people participated in an orderly and peaceful march at Fessenheim. Then in July an antinuclear festival attracted 15,000 people to the Ain region near Bugey. The starting point of visible nuclear opposition, the festival was essentially a counterculture revival of the themes of May 1968. Some months later the same group organized a "camp-in." Forty people squatted on the proposed power plant site at Bugey for fifty days, and then, with 2,000 supporters, they marched the 40 km to Lyon. This was the only large French site demonstration before 1975. The public was generally unaware of nuclear policy, and the activists focused on providing information about nuclear power and raising public consciousness about its risks.

In December 1971 activists from several countries met in Strasbourg to discuss the tactics of nuclear opposition, and at this international meeting two quite different views emerged. Some ecologists opted for radical action to arouse public opinion; others preferred to use legal channels to influence directly public officials. At this early stage these groups could coordinate their activities; later, as mass demonstrations over specific site decisions increased, their paths diverged.

Some of the more radical activists hoped to stop the whole program immediately: influenced by surrealist and situationist politics, they worked through symbolic propaganda in the counterculture press (*Charlie Hebdo* and *La Gueule ouverte*) and through theatrical techniques. In February 1972 the Paris antinuclear committee rented balcony seats at the opera during a benefit for cancer research and scattered tracts, signed by Jean Rostand and Theodore Monod, among the audience: "You're asked to give money for the battle against cancer while the government is encouraging the growth of cancer through its nuclear program." In 1974 the Parisian branch of Les Amis de la Terre organized an enormous bicycle demonstration involving more than 10,000 people.

These early uncoordinated activities served their purpose. When the government announced its all-nuclear energy program in 1974, and Michel d'Ornano, minister of industry, distributed a list of potential nuclear power plant sites, a large number of people, convinced that they had a moral obligation and a direct interest to resist nuclear expansion, were ready to demonstrate their concerns. They were supported by a number of scientists. In 1971 the eminent mathematician A. Grothendieck had created a French section of an international organization called Survivre. Together with other colleagues, mostly mathematicians, he published *Survivre et Vivre* ("To Survive and to Live") which appeared from 1970 to 1975. Scientists of the CEA, members of CFDT, also began to ask critical questions. Then, in early 1975, the Appel des 400, a petition signed by 400 scientists, was circulated by physicists at the Collège de France. This petition, suggesting that concern with nuclear power extended beyond marginal groups, was widely publicized in the national press.[2]

The growth in press coverage of the nuclear issue was exponential. A survey of articles in major newspapers found that the shift to the American light-water reactor in 1969 prompted 33 articles; the 1974 Messmer plan to expand the nuclear program, 69 articles; and the 1975 parliamentary debates on this issue, 107 articles, almost all focusing on the limited political influence on nuclear decisions.[3]

As press coverage increased, the substance of the articles changed. Until 1973 the press focused on technical and economic questions, then it turned to environmental issues. Political themes, related to decision making, began to prevail in 1975. *Figaro,* for example, ran a series called "The Great Debate," that emphasized the inability of traditional democratic institutions to handle nuclear decisions. *Le Monde* commented, "If the citizens go to the streets, organize referenda, refuse to pay their electricity bills, and draw EDF to the courts, it is because they have no other means to be heard."[4]

In France, where it is virtually impossible to influence policy without access to the highest political level, nuclear protest began with demonstrations. However, in Germany, where power is divided among different levels of administration, antinuclear activists initially raised their objections through public hearings and the courts. In Mühlheim-Karlich in the

Rhine valley, in the *Land* Rhineland-Palatinate, a teacher, Helga Vohwinkel, began an exhausting personal campaign to stop the construction of
a power plant in 1972. First she uncovered an expert report suggesting the
inadequacy of certain safety precautions. Then she organized a small citizen initiative and collected 10,000 signatures for a collective objection to
be submitted during the formal procedures. The utilities responded by
placing the proposed AEG plant with a similar Siemens plant (the two
companies are linked in the KWU). Vohwinkel maintained her objections
and began to investigate the informal relationships between the utilities
and the licensing authorities. After a battle lasting nearly five years, an
administrative court suspended the project because of collusion between
promoter and licensing authorities. This case was almost entirely the result
of an individual action.[5] Its outcome reflected a social climate increasingly
critical of the nuclear industry and the licensing authorities.

Between 1970 and 1985 the government planned to increase energy production in the Rhine valley (from Basel to the Ruhr) from 2,590 to 34,740
Mw.[6] this was part of the regional development program that would turn
this rural area into the most industrialized section of Europe—a new
"Ruhr" with industries and highways planned for both sides of the Rhine.
These plans faced active opposition. Citizen initiatives (groups) were re-
gionally and internationally coordinated through the Action Rhine Valley
to oppose industrialization plans for the upper Rhine valley in Switzerland, France, and Germany. Twenty-one German and French groups belonged to a joint coordinating committee that had grown out of a collaboration to oppose a chemical plant in Marckolsheim on the French side
of the Rhine. A German firm had originally planned to build this
plant in Germany, but local authorities denied a construction license. The
French, wanting foreign investment, accepted it, but French and German
groups together occupied the site and stopped the project. This incident
established solidarity among citizen initiatives on both sides of the Rhine.
Just as the first antinuclear groups were organizing on the French side of
the river, the government of Baden-Württemberg announced its plan to
build a nuclear power plant in nearby Breisach. Citizen initiatives mobilized throughout the region. They agreed on the importance of long-term
regional coordination and on a slogan: "Better active today than radioactive tomorrow!" Their tactics were to convince the farmers in the area that
nuclear power would destroy their economic base.

In September the farmers of the region organized tractor demonstrations. Then more than 60,000 people, in some villages up to 90 percent of

the adult population, signed a petition against the construction. Prime minister of the *Land,* Hans Filbinger (CDU), contributed to the growing local protest when he attributed the opposition to the influence of external extremists. Protest accelerated when a citizen initiative uncovered an emergency plan for dealing with accidents at a research reactor in Karlsruhe. It called for using the police to detain the local population within the area in order to limit the spread of contamination. All of a sudden the warnings of the activists seemed a potential reality.

After a dramatic public hearing in the fall of 1972, the government abandoned the Breisach project. It did not, however, abandon its plan to build nuclear power plants in this region, only moving the site a few kilometers away to Wyhl.

A Turning Point in Wyhl

In early 1973 the government of the *Land* chose Wyhl as the new site, and the mayor and municipal council agreed in principle to sell the necessary acreage to the utilities.[7] Citizen initiatives immediately formed in the surrounding villages. To avoid a protest similar to the one in Breisach, the utilities organized an information program and issued a monthly brochure on nuclear safety and on energy needs: "Without nuclear energy, the lights will go out in the mid-1980s." The population polarized over the issue as a citizen initiative formed to support nuclear power.

When the *Land* administration started licensing procedures in 1974, the opposition groups took action. A 6-km-long procession, including thousands of wine growers and farmers from the whole region, marched to Wyhl.

The citizen initiatives used a donation from the local hunting association to hire several scientists to criticize the documents. The scientists concentrated on the climatic effect of cooling towers and the consequences for local agriculture. They helped to build a weather observation station to monitor the microclimate. This focus on specific local climatic impacts added to the standard arguments about nuclear safety and had a powerful political effect. When in 1974 the administrative notice of the plant appeared in the official journal, 8 communes, 50 associations, and 330 individuals lodged formal objections. Also 90,000 people signed a collective objection formulated by the citizen initiatives.

The hearings held on June 9 and 10, 1974, shaped the subsequent conflict. The citizen initiatives contended that the chairman of the hearing, an administrator from the ministry of economy in the *Land,* was arbitrary and unfair, that utility experts dominated the procedure while the farmers were dismissed as lacking competence. The initiatives boycotted the hearing, and the BBU, as well as a CDU member of the *Land* parliament, asked for a new procedure.

Despite local protest the *Land* government would not consider changing its plans; no official could be found to discuss the matter with a delegation brought together by the initiatives in Stuttgart. When the *Land* parliament finally discussed the Wyhl case in November 1974, only 25 of its 120 members participated. Local SPD and FDP groups and even some CDU members of the municipal councils supported the opposition. But in the *Land* parliament the CDU fraction had to maintain loyalty to the *Land* government, and SPD and FDP parliamentarians had to avoid embarrassing the government in Bonn.

The Protestant church was the only major organization to oppose unequivocally the project. Less concerned with the hazards of nuclear technology than the undemocratic nature of the procedures, the church stressed the need of public participation in planning such projects and condemned attempts by officials to criminalize antinuclear activists.

As hostility increased in late 1974, the mayor of Wyhl scheduled a local referendum on whether the municipality should sell land to the utilities. The referendum inspired an active campaign. The prime minister of the *Land* sent a personal letter to all households favoring the project, and landowners were led to believe that a negative vote would lead to expropriation. Meanwhile, citizen initiatives convinced prominent scientists and a large part of the academic community of the nearby city Freiburg to campaign against the nuclear plant. Twenty-two scientists signed a statement warning of its potential ecological and health hazards. The referendum on January 12, 1975, brought out 92 percent of the population: 55 percent voted in favor of selling, 43.2 percent against.

The licensing authorities acted promptly: On January 22, the first partial license was granted allowing construction to begin. But then the neighboring municipalities responded. Some 3,000 people met in Weisweil, and 4 municipalities decided to take the issue to the administrative court in Freiburg. This court advised suspending construction until it could evaluate the complaint, but the utilities proceeded. At this point 150 protestors occupied the plant site. The *Land* authorities labeled them left-wing extremists, and the police moved in, clearing the site and arresting about 50 people. A few days later, in February 1975, 20,000 people appeared at a demonstration that culminated in a massive site occupation. The police withdrew, and Wyhl became a mecca for antinuclear activists from all over Germany, France, and Switzerland.

In March the administrative court blocked the first partial construction license. The utilities appealed the decision, but construction stopped. Meanwhile the popular mobilization around Wyhl continued; mass meetings at the occupied site involved thousands of people, and political authorities were increasingly concerned about the inevitable violence should the police try to expel the occupants.

By this time the tactics were reduced to erosion. The utilities announced an increase in electricity prices. When the electricity failed during a popular international soccer game, it was dramatized as an example of future electricity shortages. The minister of economy of the *Land* inveighed against a television program sympathetic to antinuclear actions. The utility sued people involved in the site occupation for damages up to 46,500 DM. The citizen initiatives, meanwhile, strengthened their organization and their collaboration with French groups who were fighting the Fessenheim plant and Swiss groups who were occupying a proposed plant site in Kaiseraugst.

The citizen initiatives also created a "popular university" at Wyhl. With the help of academics from Freiburg, they organized sixty-three courses, bringing in scientists, including several known for their pronuclear opinions. Erhard Eppler, the president of the SPD *Land* organization, lectured on "Wyhl and Democracy." Lectures and discussions included such topics as environmental protection, the police state, the right to resist and defend oneself, and the principles of democracy.

After more than four months of such activity a compromise was reached: the authorities agreed to consider the expertise provided by citizen initiatives concerning the impact of the plant and to withdraw any claims for damages from the occupation. They guaranteed that construction would start again only after a final court decision. The initiatives then left the site in January 1976. After a long trial in March 1977 the administrative court finally decided to ban construction. The judgment turned on a single technical point: insufficient provision had been made for a concrete containment vessel in the reactor design.

Wyhl was a turning point for the antinuclear movement. It showed that a local population could maintain a high level of militancy, that endurance and good legal and technical advice could bring success in the courts, that well-informed citizen initiatives could use the issue to raise fundamental political questions about democratic procedures, and finally that the tactics of site occupation, with its implicit threat of violence, could create conflict among politicians and uncertainty within the nuclear establishment. Wyhl thus had both national and international importance, encouraging antinuclear activists throughout Europe and in the United States as well.[8] But while the antinuclear movement drew encouragement from the event, political authorities drew their own conclusion: under no circumstances would another site occupation be allowed to take place. While activists saw site occupations as their most important weapon, the authorities were determined to prevent such actions at any price. Later conflicts were shaped by these contradictory perceptions, and violence was the price. After Wyhl the antinuclear movement turned toward a pattern of direct confrontation.

A Quasi-Civil War?

In 1973 the utility announced its plans to site several nuclear power plants in Brokdorf—a region that was a leisure center for the Hamburg urban area. In fact the plans had existed since 1971, and an agreement had already been reached between the utility Nordwestdeutsche Kraftwerke (NWK) and the government of Schleswig-Holstein.[9] In November 1973 the president of the Brokdorf tourist association began to mobilize resistance. She sent a questionnaire to everyone in Brokdorf and Wewelsfleth, and two-thirds of the people responded: 784 (75 percent) against the power plant, 202 (20 percent) in favor; 500 people did not respond. She also helped to organize a regional citizen initiative, and the mayor of Brokdorf was one of its founding members. The mayor, however, changed his mind when the utility guaranteed to finance several construction projects in his municipality. Forced to deal with the questionnaire, he argued that the 500 people who had failed to respond actually favored the plant. On this basis he recalculated the results, announcing that less than half of the population opposed the plant.

During the following months prominent politicians from the CDU tried to reduce local opposition in the villages by labeling the protesters extremist and communist. In a local newspaper, Kai Uwe von Hassel, an ex-CDU minister of defense, criticized "the upright citizens of the Wilstermarsch who accept being put before the wheelbarrow of communists." When the mayor of Wewelsfleth called for public participation in the planning process, officials of NWK asked him if he had "communist propensities to establish a soviet republic." [10]

In August 1974 NWK officially requested a construction license. The local citizen initiative responded with a petition signed by 31,000 people; other citizen initiatives supported the objection in the name of their 40,000 members; and 160 people objected on an individual basis.

In November 1974 the public hearing witnessed antinuclear arguments presented by farmers and fishermen, experts from Bremen University, and the American activist Arthur Tamplin. They raised questions about the adequacy of the government's argument and the completeness of the dossier. When a Liberal member of parliament supported these concerns, the CDU prime minister of the *Land*, Gerhard Stoltenberg, insisted that the hearing be legally terminated.

The citizen initiative continued to develop contacts with other regional and national associations, in particular the Weltbund zum Schutz des Lebens (WSL) and the *Bund*. It also organized activities to expand its constituency. Hundreds of people attended a May fire dance. A street theater group created a satire on the nuclear controversy called "To Be or Not to Be" and played it in every village in the region. Where street theater was forbidden, the Protestant church provided the necessary space. As in Wyhl the tactics of protest combined legal action and site occupations.

The utility responded by supporting a pronuclear citizen initiative called Intelligence Forward. It hired a public relations agency to plan its activities and opened an information center in Brokdorf. (Between September 1975 and January 1976 only 250 people used it.) As protest increased—evident in the thousands of objections filed during licensing procedures—the government decided to visibly demonstrate its strength. The Hamburg police confiscated 10,000 copies of a tract providing tactical information. Police, armed and with dogs, explained their presence: "We are here to protect you against radical elements."

NWK coordinated its construction plans with the licensing authorities; material and machines arrived at Brokdorf on the night before officials planned to grant the first license. By morning the police had built a trench and a wall around the construction area. On October 30, 1976, 8,000 people marched onto the site, and some 500 occupied the construction area. After dark the police expelled them with horses, high-pressure water hoses, and chemical mace. Several people were wounded; 52 were arrested. The reaction from all over Germany was overwhelming. The next day 4,000 people participated in a silent march protesting police violence, and the citizen initiatives called for another demonstration on November 13. Meanwhile the construction area became a fortress, the water trench enlarged, the walls reinforced.

The day before the demonstration police closed the roads to Brokdorf, and demonstrators—some 30,000 to 45,000 of them—marched several kilometers to the site. On November 11, 1976, a Protestant minister held an open-air Mass. All the leading ecologists were there to express the solidarity of the movement. A battle, often described as civil war, began when demonstrators tried to occupy the site. The demonstrators broke through the wall around the construction area. But the police were prepared. In the evening, when the demonstrators began to leave, police in helicopters showered them with mace. More than 500 people were wounded and 100 arrested.

Brokdorf was the most violent confrontation to date, and it had a complicated and divisive effect on the antinuclear movement. On the one hand, thousands of people expressed their willingness to cooperate in future actions. Letters came from environmental associations, Catholic and Protestant youth organizations, university student parliaments, professional societies, twenty local SPD organizations, several FDP local youth groups, and thirty suborganizations of the DGB. On the other hand, the media blamed the violence on the demonstrators, and many viewed the antinuclear movement as a threat to public order. Moreover even among activists the increasing use of violence by the police deterred further action.

Growing divisions within the movement emerged in discussions about the next demonstration planned for February 19, 1977. Some groups wanted to gather everybody in Germany and in foreign countries once more in Brokdorf. Others, mainly from the local area, were afraid to create a situation close to civil war, and they voted for a meeting in the nearby town of Itzehoe.

The press campaign against the extremists intensified, although extremist groups were only a small minority of the thousands mobilized by the initiatives. Prime Minister Gerhard Stoltenberg stated that the East German government had spent several thousand DM to infiltrate the movement.[11] Officials reported the existence of secret protocols in which radical groups claimed they possessed enough welding materials, weapons, and catapults to attack the police armory.[12]

Just before the February demonstration the administrative court decided to withdraw the Brokdorf license on the grounds that inadequate provision had been made for waste disposal. By linking construction to the solution of the waste disposal problem, this court decision implied a de facto moratorium that would significantly slow down the German nuclear program. Tensions abated, and, while the demonstrations held at both Brokdorf and Itzehoe drew some 60,000 people, they were peaceful.

Citizen initiatives also organized opposition at three other sites in Esensham, Ohu, and Grohnde. The sequence of events at Grohnde resembled the pattern at Wyhl and Brokdorf. At the nuclear hearings officials failed to respond to some 12,000 objections. The public was excluded from the water pollution hearings where the outcome was clearly determined months earlier by a *Land* official who stated that the nuclear project was consistent with the water pollution laws. Once public authorities granted the first construction license, police surrounded the area with a fence that critics called the "Berlin wall" or a "concentration camp border." At the first demonstration some 1,000 antinuclear activists surprised the police with flowers and carnival confetti and then occupied the site for two hours. They withdrew voluntarily when faced with hostility from local citizen initiatives. Divided over the issue of violence, the activists finally agreed to organize two separate actions on March 19. Local groups of conservative ecologists would meet in a village neighboring Grohnde; the radical groups, mainly from larger cities, would demonstrate at the site.

The administrative response to these events was quite different than in Brokdorf. Before the demonstration the Lower Saxony government released no information and asked the media to maintain silence, hoping that silence would reduce the possibility of a theatrical event. But nearly 20,000 people came to Grohnde. The demonstrators had learned from experience. They brought helmets, face masks, and kerchiefs to protect themselves against police bludgeons and chemical mace; they built aluminum kites to disturb the police telecommunication system and hamper helicopter attacks; and they organized first-aid and transportation services. The actions were well organized and directed by walkie-talkies.[13]

The police in turn reinforced their mounted divisions and antiriot technology. Around 1,600 demonstrators and several policemen were wounded. Prime Minister Ernst Albrecht (CDU) called the participants criminals, and the newspaper estimated that more demonstrators than ever before were willing to use violence: conservative newspapers detected

"5,000 lovers of chaos" *(Morgenpost)* and a "Civil war army of 3,000–5,000 members" *(Die Welt)*; liberal newspapers estimated some 1,000 extreme radicals among the demonstrators *(Frankfurter Rundschau)*. It was generally reported as the worst violence in the Federal Republic since its founding. The polarization over the issue of violence increased, and the citizen initiatives began to realize the impossibility of success through such quasi-military confrontations.

One more major antinuclear demonstration took place at the site of the German fast-breeder, Kalkar.[14] Here the social-liberal government of North Rhine-Westphalia tried a new approach. The minister of the interior confirmed the right to demonstrate but emphasized his responsibility to maintain public order; he would impose all the necessary constraints to guarantee a peaceful demonstration. Police forbade demonstrators from wearing protective helmets or masks and from carrying any instruments that could possibly be used as weapons. Loudspeaker trucks, red cross signs, ambulances were all excluded.

The day before the demonstration, on September 24, 1977, police controlled all the major roads and highways approaching Kalkar. They pressured bus rental companies to avoid doing business with antinuclear groups. Yet a thousand buses descended on Kalkar from throughout the Federal Republic. More than 10,000 Dutch and 1,000 French activists came as well.

The police and border control tried to delay their arrival, stopping and inspecting buses as many as ten times on route to Kalkar. Police in helicopters stopped a train several kilometers from the city and questioned all passengers. Helmets, masks, and all possible weapons were confiscated. Police even took away bottles of lemon juice, useful for counteracting the effects of tear gas. German officials tried to mobilize xenophobic feelings by pointing to the participation of extreme French groups; 400 French citizens were denied entry to the Federal Republic.

Fifty thousand people arrived at Kalkar, but the constraints and controls had made a mockery of the right of demonstration. Some days later the BBU declared officially that it no longer considered mass demonstrations and site occupations the major means to achieve ecological goals.

Despite the rhetoric of civil war that pervaded so many descriptions of events in Germany, it was clear that potential sympathizers of the antinuclear movement would never support a violent struggle over this issue, leaving the most radical activists increasingly isolated.

An Attack on the Symbols of Power

Braud-et-Saint Louis in the Gironde is a small village of 1,200 people, mostly wine growers, fishermen, and farmers who cultivate the *marais*—the bog where EDF planned to locate a nuclear power plant. In 1974 EDF announced that construction would soon begin on four reactors (two 900 Mw plants and two 1,300 Mw plants) to be completed between 1980 and 1985. The utility acquired 250 hectares and began work. Local farmers had already formed a group to protect the *marais;* the village had recently suffered through a wine-growing crisis and could not tolerate further dislocation.

Braud's farmers were backed by a regional environmental group, Societé pour l'Etude, la Protection et l'Aménagement de la Nature dans le Sud-Ouest (SEPANSO), whose interest in nuclear policy had first been aroused by an earlier plan to bury waste in the gulf of Gascon. SEPANSO called for a public debate, but EDF refused: "A public discussion would not be justified, for there already exist municipal officers and deputies. Moreover this procedure does not permit a dialogue and an objective and serious debate." [15] Then at the public inquiry SEPANSO mobilized antinuclear scientists from the three departments near the site to provide technical criticism and gathered 26,000 signatures on a petition against the plant.

In January 1975 the regional assembly approved the site, but construction began before the permit was actually in hand. This further radicalized the opposition: a bomb exploded in the house of the conseil général, some EDF trucks were destroyed, and farmers occupied an EDF building. Police broke up the occupation and the courts convicted twelve farmers under the antidemonstration laws that had been promulgated after May 1968. This, however, only increased protest; activists, including students, seasonal farmworkers, teachers, and environmentalists, set up camp on a farm adjacent to the construction site. On July 9 CRS, the national antiriot police, intervened.

Everyone in the village of Braud was involved, and the issue polarized the population. Storekeepers wanted the development that comes with industrialization; farmers were solidly against it. But even those favoring the construction complained about lack of information. The mayor of Royan asserted that the local population was completely ignorant of the consequences of the project. The municipal council of Braud voted for a five-year moratorium, but its mayor, Kleber Marsaud, who had held this office for twenty-seven years, admitted, "I have no power to apply a moratorium. I am only a small country mayor." [16] Construction went on.

Two other major protests took place in 1975, one at Flamanville in Normandy, the other at Fessenheim in Alsace, both earmarked as major nuclear centers. Flamanville is a town of 1,400 people on an abandoned iron mine. In 1974 EDF announced its plans to build four plants of 1,200 Mw each on 120 hectares of land. High-tension wire would transmit the electricity generated in Flamanville to Brittany and Paris.

The socialist mayor and the municipal council of Flamanville welcomed EDF's plan; it was an opportunity to bring employment and much needed prosperity to the area.[17] Construction would employ 2,000 people for six years and bring an estimated 40 million fr to a town whose communal budget was only 450,000 fr. But a number of local people feared the disruption likely to follow such rapid change. They also questioned the decision-making procedure; information was published twenty days after the municipal council had already voted to accept the plan.

In December 1974 thirty environmentalists met and raised several questions: What would happen to the 2,000 employees after the construction was over? What would be the future of the local fishing business? What was the significance of the size of the land expropriated by EDF? Few local people had thought about the size and scope of this project, and on January 3 a hundred people attended a second meeting, forming an antinuclear committee. Five days later a delegation from the conseil général of the department arrived with some engineers to survey the site and discovered the road to the village barred. On January 25, 250 local residents attended a meeting at which EDF engineers and local government officials debated environmentalists for over four hours.

As in Braud the village polarized. Blue-collar workers and shopkeepers supported the plan: "To live is to work." Farmers and fishermen feared its implications for their occupations. Environmentalists warned of degenerate descendants; advocates called the critics "Maoists," "Jehovah's Witnesses," or just "politicians." The local curé favored the plant: "The atom is a part of the creation after all." Others talked of bread, wine, and swimming pools. (EDF had organized a visit of community leaders to the model plant in St. Laurent-des-Eaux with its three swimming pools.)

Incidents increased; tires were slashed; the mayor received a bullet in the mail. A local butter factory, concerned that the association with radioactivity would affect sales, changed its brand name. Faced by growing pressure, the mayor of Flamanville called a referendum on April 8, 1975: 435

people voted for the plant, 248 against. However, the following Sunday thousands of environmentalists once again marched to the site.

The battle continued over the next year. Twenty-five local antinuclear organizations formed a regional committee to help distribute information. Farmers organized themselves to buy up any land about to be sold to EDF. Meanwhile a strike by technicians at the nearby La 'Hague reprocessing plant over working conditions encouraged opposition, indicating that technicians themselves were concerned about radiation risks.

At the end of 1976 the government held a public inquiry in five nearby communes. That the required environmental impact statement was not in the dossier added to local suspicion, and in a local referendum (but one with no judicial authority) 65 percent voted against the plant. Nevertheless, on February 8, 1977, EDF announced that the public inquiry had resulted in a favorable report and work on the site began. Twice protesters forced the workers to evacuate the site, and by March there was open confrontation as police and protesters alternately occupied the construction area.

Meanwhile another major dispute was taking place in Fessenheim, an Alsatian village of 900 people in a rural area rapidly undergoing industrialization.[18] Few farmers are left in Fessenheim, and half the working people are employed in industry. EDF planned two reactors along the banks of the Rhine on land that it had acquired in the 1950s. In late 1974 EDF began a public relations campaign with tours of model power plants, films, and documents to convince local government elites that nuclear power would bring prosperity to the region. In response a regional group, Comité de Sauvegarde de Fessenheim et de la Pleine du Rhin (CSFR), comprised mainly of students and teachers from outside the village, organized its own information campaign. Indeed the local people were saturated if not glutted with contradictory advice, and they reacted with confusion, doubt, and often withdrawal. Some felt it useless to try to understand such a complicated problem, especially when even those in power could not agree. Those with opinions believed they could not affect a national decision. More important, in this small community it was crucial to avoid conflict and polarization.

As in other villages local farmers welcomed environmentalists, but the working class people remained "behind their curtains," avoiding discussion with CSFR members who came from a different social class and cultural background. In this population, in transition between rural and industrial economy, the environmental ideology was alien, but the idea of confronting a national decision was even more so. Thus the response took a curious form. One Friday CSFR announced it would distribute information about nuclear power door-to-door on the following Monday. In church on Sunday the priest warned that Jehovah's Witnesses were coming around on Monday. Doors were therefore barred.

In the spring of 1975 two bombs, placed by someone who was obviously knowledgeable about the reactor, exploded at the site. A group of sympathizers of the German Red Army fraction claimed responsibility.[19] The bomb caused little damage but considerable public concern when the press announced the act as one of terrorism. It also concerned environmentalists, who were mostly committed to peaceful protest:

This means of protest is not in accord with our methods . . . but if democracy had functioned in the case of Fessenheim, . . . a real debate would have taken place, and we could have together chosen our future.[20]

Opposition progressed toward increased attacks on property—the bombing of buildings and the burning of official documents. The violence, for the most part, was directed toward things—the symbols of the nuclear establishment. Similar events took place at other nuclear sites—at Le Pellerin and Port La Nouvelle. This phase of antinuclear activism in France culminated in the largest demonstration of all—at the site of the Super Phoenix breeder reactor in Creys-Malville in July 1977.

In 1974 EDF requested a construction license for the 5,000 Mw fast-breeder reactor Super Phoenix at Creys-Malville and conducted a public inquiry.[21] Ecologists sought to stop the project through court action and blocked the first DUP procedure because it did not conform to legal requirements. EDF started work on the construction area before completion of the second procedure in November 1974. This time the courts rejected the ecologists' complaint. The government, committed to this high-prestige project and allied with Italian and German utilities in developing it, paid little attention to the rising protest. Faced with this challenge, antinuclear activists used increasingly disruptive tactics.

The first mass demonstration took place in the summer of 1976. The ecologists established a pirate radio station, "Radio-active." This was quickly banned as a violation of the French public radio monopoly legislation. Some days before the demonstration the prefect warned the local population that ecologists could bring violence to their communities, and several municipalities prohibited camping and traffic to discourage participation. Nevertheless, some 15,000 French, German, and Swiss ecologists gathered and stayed for a week. The government mobilized five divisions of the CRS (the antiriot police), and they evacuated the site: four demonstrators were hospitalized. Three municipal councils protested against the police brutality. The prefect who had ordered the police action responded: "We warn, and then we clean up."

In September 1976 twenty people penetrated the prefecture at Grenoble and a few days later presented stolen documents to the press; these included the Orsec-Rad emergency evacuation plan, circulated as proof of the dangers embodied in the project. Police arrested two members of the local antinuclear group and charged them with theft, complicity, concealment, and irregular entry. The event, however, further increased public concern. The regional conseil général of Isère, in which socialists hold a majority, organized a debate, inviting legislators, representatives of ecology associations, EDF, and CEA. The conseil finally voted to suspend construction. A few months later the conseil général of Savoy and the council of Geneva in Switzerland adopted the same position. But the government simply continued to ignore these manifestations of public opposition.

Over the next year antinuclear demonstrations took place in cities close to the site, Grenoble, Geneva, Lyon, and Valence. The regional organizations of the CFDT, the Fédération de l'Education Nationale (FEN), the Socialist party, and the PSU took firm positions against the project. Mayors and members of municipal councils created an association called Local Elected Representatives against Super Phoenix. Scientists intervened in the debate: 1,300 engineers, physicists, and technicians from the European Center of Nuclear Research (CERN) in Geneva sent a petition to the French, Italian, and German governments, and 504 scientists from the Rhone-Alps region sent a petition to the president.

When these efforts failed to provoke a response, ecologists called for a major demonstration on the construction site at the end of July 1977. Police arrived weeks before this event, determined to prevent a site occupation, but 60,000 people appeared. The first day was relatively peaceful, but the prefect in charge of the police made an inflammatory speech on the radio that changed the mood. Referring to a few drunken German demonstrators who had penetrated the city hall of Morestel, he dramatically announced: "For the second time Morestel is occupied by the Germans." Comparing the antinuclear activists with Nazi troops added to the tension. The next day clashes between demonstrators and the police could no longer be avoided.

Several hundred demonstrators were wounded, and one of the victims, Vital Michalon, died the next day. The medical service of the antinuclear groups that treated him certified that he died of a pulmonary injury caused directly by a blast from a grenade. Later officials used other medical reports to claim a heart attack. Michalon became the first political martyr of the movement.

The escalation of the nuclear protest in the two countries had different roots. In Germany critics felt the administration had broken its own rules, and therefore anger was a powerful force in legitimizing even violent resistance. The citizen initiatives compared the use of the police to suppress antinuclear demonstrations to Nazi politics. A women's initiative in Baden articulated this comparison in an open letter to Prime Minister Hans Filbinger.[22]

Some German critics claimed that the implications of nuclear energy for a police state constitute a threat to the constitutional order, and they referred to the constitutional provisions allowing citizens to resist actively such threats.[23]

In the French context such moral, historical, and legalistic arguments played a less important role. French activists thought rather in immediate political terms of power and social control. Disillusioned with official channels of influence, they used more radical tactics as a means to increase the political consciousness of the public and create uncertainty among nuclear authorities.

Violence in these circumstances took somewhat different forms. In France until Creys-Malville activists directed violent actions against the symbols of the nuclear establishment—the offices of EDF and Framatome, the official documents, and the dossiers. In Germany violence grew out of collective action that activists themselves described as a war on the battlefields of nuclear sites. The demonstrators believed they could win these violent confrontations and thereby change government policy. They clearly overestimated their capacities and underestimated the willingness of government to use legitimate violence to put down the demonstrations.

Violence polarized attitudes toward nuclear power and reduced the space for negotiation and compromise. The preparation of mass demonstrations had sapped the energies of the activists. In Germany a certain success came from legal actions, as administrative courts had delayed construction at Mühlbach-Kärlich, Wyhl, Brokdorf, Grohnde, and Kalkar. In France, however, the exhausting mobilizations of 1976 and 1977 produced resignation, as activists felt that by trying to stop everything they had achieved nothing at all.

The self-immolation of a German activist, Hartmut Grundler, one day before the SPD was to decide its energy policy, painfully dramatized the mood after the demonstrations of 1976 and 1977. Grundler's last letter, directed to Chancellor Helmut Schmidt, to SPD party chairman Willy Brandt, and to the citizen initiatives, was a moral plea: it asked Schmidt and Brandt to reveal the pressures of the nuclear industry on government and the lies of the nuclear lobby; it begged them to change policy or to resign; and it implored the citizen initiatives to persist in nonviolent tactics in the name of humanist ideals.

6 Tactical Adaptations

When the first bombs exploded at Fessenheim, those claiming responsibility justified their action with a slogan from the student movement: "Yes to violence against property—No to violence against people!"

We took all possible precautions to avoid threatening human life. That is our contribution to the antinuclear struggle. . . . The capitalists no longer hesitate to extend their traditional genocide (wars, factories, prisons) into a much more radical form of genocide represented by nuclear power. . . . Our enemies are . . . multinational corporations. . . . We hope that all those who approve our actions will imitate us. Eco-sabotage has started.[1]

Most local citizen groups on both the French and German sides of the Rhine dissociated themselves from the violence at Fessenheim, although they emphasized that this should not be interpreted as sympathy for public authorities. Les Amis de la Terre, however, approved the action so long as it respected human life:

[We] express solidarity with those responsible for the bombing in Fessenheim. We deeply admire the skill of their action: it was efficient; no people were hurt; it was particularly opportune. . . . Resistance by any means that do not threaten human life is a moral act.[2]

The reaction in Germany reflected the political climate shaped by years of terrorist activities. No political group could afford to take a position that distinguished violence against property from violence against people. Thus, except for the most radical student groups, German organizations condemned the bombing, while blaming the undemocratic behavior of the nuclear industry and public authorities for the despair that motivates such events.[3]

In both countries violence as a source of divisiveness within the antinuclear movement forced the activists to develop more diverse strategies, including electoral participation, legal action, and new forms of civil disobedience.

The Divisiveness of Violence

The debate over violence sharpened when lives were at stake: How much escalation should activists encourage when they know that they bear responsibility for thousands of demonstrators? Can a social movement survive if the escalation of violence causes major splits over appropriate tactics?

These questions assumed increasing importance in Germany after the Brokdorf demonstrations. In both Brokdorf and Grohnde the organization of demonstrations caused major divisions in the movement as some groups opted to avoid police clashes at any price, while others hoped for some degree of calculated conflict. As one participant explained: "The local farmer must overcome a mental barrier before he can accept an illegal site occupation that would disrupt the peace. For him the risks are much higher than for a student." Addressing the danger of a split between its young radical student constituency and local people, a BBU brochure on tactics instructs its members: "Action is not tourism! The local people are the first concerned. They have to bear the main burden of resistance. Help is welcome, but only if unselfish." Theoretical reflections about violence tend to dissolve in the heat of a demonstration:

I had always approved of site occupations. . . . All of a sudden faced with reality, I hesitate. Several objections come to mind: it does not make sense, we will not be able to stay; I am not adequately dressed. What to do: to occupy the site now or to go home at 7 p.m.? What if I get arrested, hurt, wounded. . . . I realize that the right to resist is more difficult to implement than I had thought, and obviously, hundreds around me think the same thing—there are mental barriers even against minor violence such as abolishing a wall . . . around me people start to walk in the same direction. Finally, I follow and cross the water trench.[4]

The dynamics of a demonstration, the immediacy of the action, its dangers, and its crowds carry participants along. But the provocative behavior of even a few demonstrators may turn people away:

I was terribly afraid standing near Maoist activists who threw a molotov cocktail into a high-pressure water truck and flames came out of it. These people yelled and screamed enthusiastically! Suddenly I remember that's the way a policeman died in a police car in Frankfurt a short time ago. When I asked these people, "Do you want to kill the cops?" they pushed me away. And they replied, ". . . Should Nazi pigs survive? They are the murderers!" [5]

Demonstrations were especially destructive in Germany. They repelled local people who saw their villages transformed into battlefields. They put off many of the scientists who provided the movement with expertise and moral legitimation. They cut the movement off from important sources of support—from intellectuals, from religious groups, and from others committed to nonviolent tactics.

One issue of the BBU's newsmagazine was almost exclusively devoted to nonviolent political tactics and their history. Articles on Mahatma Gandhi reconstruct an ideological and historical tradition linking Gandhi's philosophy of nonviolence to antinuclear and ecological activism.[6] The mainstream of the antinuclear movement chose to defend itself against accusations of violence by establishing historical continuity among all nonviolent movements, with Gandhi as a spiritual father.

But within the movement a small active minority felt that only violent resistance would break the commitment to nuclear energy. The radicalization of these more extreme activists increased, as authorities marginalized their critics.

In the French ecology movement the most significant split took place between a group of radical ecologists and the emerging autonomous movement. The ecologists are organized around *La Gueule ouverte*. Committed by principle to nonviolence, they believe that in the context of power relationships in society today people must invent nonviolent techniques of resistance: "The people are not a weapon." Beyond their moral objections to violence ecologists assess the question of tactics in terms of political realities; the movement cannot expect to demolish the state in a direct confrontation. It can, however, create new political cleavages and bring new issues to the policy agenda. Thus for most ecology groups the adjective "radical" points more to their analysis than to their behavior, which often focuses more on raising public consciousness than on political action.

They are, however, often joined by groups with quite different agendas. The autonomists for example are a small group of disaffected youth who engage in disruptive and violent action, often for its own sake. They justify their actions as counterviolence—a response to the state's monopoly of violence. Autonomists define their goals less by reflection on the historical necessity of a revolution than by immediate needs and desires.[7] With no vision of a new society, their struggle is an end in itself: everything is to be achieved here and now. The ecologists' plea for nonviolence has little influence on the autonomists, and the two groups have been unable to come to terms. Indeed the threat of violence unconstrained by a vision of the future has forced antinuclear activists to avoid situations where the autonomists could become involved.

After the violent demonstrations of 1977 the major organizations in both France and Germany decided to diversify their tactics. This did not preclude the use of demonstrations; indeed the Gorleben International Review in the spring of 1979 revived activism and generated the largest mass demonstration of all. But this event took place in a much more diversified social context. By then antinuclear activists were involved in electoral politics and legal actions as well as protest, and they had begun to focus their efforts on a few selected targets—the key projects in the nuclear program.

Oscillations from Civil Disobedience to Electoral Participation

After two years of active mobilization mass demonstrations essentially ceased between 1977 and 1979. Some observers concluded that this was the end of the antinuclear movement. But ecology publications proliferated, and citizen initiatives grew. Activists participated in electoral politics. They worked through the courts to block specific siting decisions. They traveled to maintain coordination among local and regional groups and to develop international cooperation. They invented new forms of civil disobedience.

Antinuclear groups also sought alliances with other associations. In France Les Amis de la Terre and SOS-Environnement established relations with the consumer defense federation, Fédération Française des Consommateurs (FFC), which devoted one issue of its magazine *Que choisir* to the nuclear debate and another to alternative energy sources. The leadership of the movement also tried to build up the political skills of local groups. BBU published a guidebook reviewing the tactical principles of the movement and providing detailed practical advice on how to print a brochure or design a bumper sticker, on how to run a press conference or take a case to court, on how to lobby political parties or influence local administrations.[8] In France Pierre Samuel of Les Amis de la Terre wrote a similar guide to political action.[9] Both books encouraged nonviolent action and insisted that any confrontation that could result in violence should be avoided: "We have to be able to walk the fine line that separates adaptation to the existing system (integration) from complete rejection of it (negation)."[10]

For both BBU and various French antinuclear organizations, nonviolent tactics include civil disobedience. For example, they urged people to withhold the percentage that the utilities devote to the expansion of nuclear energy (around 10 percent) from their electricity bill. A Hamburg citizen initiative convinced several hundred families to participate. All of them paid the 10 percent deducted to a common account for the legal defense of antinuclear activists. As this practice spread, the utilities went to court. The citizen initiatives defended their action based on their moral right to resist nuclear power as a threat to survival. But the courts rejected the plea and allowed utilities to cut off electricity in cases of nonpayment. In the summer of 1979, in a similar case, a court in Stuttgart decided in favor of the protesters, refusing to allow the utilities to cut electricity.

In France citizen groups launched a national rate-withholding campaign, asking people to deduct 15 percent from EDF bills. More than 2,500 households participated. In the region of Le Pellerin workers refused to implement an EDF order to cut off the electricity in a neighborhood where people had refused to pay their bills. When EDF did cut off electricity, the citizen groups sought different ways to hassle the bureaucracy. They advised customers to forget to sign their checks, to pay a few centimes too little or too much, to pay their bills in several parts—all to disturb the computerized accounting system. These symbolic actions served to demonstrate the continuing opposition to the nuclear program, to maintain a state of perpetual uncertainty among public authorities and organizations responsible for nuclear development, and to keep the issue alive.

The most visible evidence of tactical reorientation within the movement was its increasing participation in electoral politics. In France electoral participation is often a means to bring information to the public, and candidates with no chance of success may run for office simply to get access to the media. In 1974 Les Amis de la Terre sponsored René Dumont in the presidential elections. Dumont, a well-known scientist specializing in agriculture, Third World economics, and demography, organized his campaign around such themes as the limits to growth, exploitation of nature, and problems of world population growth. Sympathetic to the Socialist party, Dumont presented these themes in an anti-imperialist, progressive ecology platform that attracted wide publicity.

The election campaign separated the traditional environmentalists from the more politicized groups: the national FFSPN did not support Dumont's candidacy while the regional and more politicized associations of Alsace and the Rhone valley did. Surprisingly Dumont captured nearly 3 percent of the vote, a sufficient number to turn the ecologists into an important element in the balance between the right and the left. But this was again divisive and confusing for the movement. Many ecologists, disillusioned by the traditional political parties, urged Dumont not to align with any one party on the second ballot. Others feared that an ecology candidate would divert votes from the left. The subsequent compromise illustrated the dilemma faced by ecologists when they enter the electoral scene; while refusing to state officially a political preference, Dumont personally declared that he would vote for Mitterand, the candidate of the united left, on the second ballot.

In another election ecologists sponsored Brice Lalonde for a seat in parliament from the fifth arrondissement of Paris. He became the first political representative from the post-May 1968 generation. Electoral participation of ecologists in the 1974 cantonal elections demonstrated that their chances on the local level were far better than on the national scene; in some areas they won more than 10 percent of the vote.

The 1977 municipal elections were the first real test for the ecology candidates, and they won seats in many municipal councils. Averaging more than 7 percent of the national vote, ecology groups became a political force that could no longer be ignored by the parties.

An analysis of these elections shows that ecologists were most successful in those rural areas where the local population had mobilized against specific industrial projects or nuclear power plants. In some small villages close to Fessenheim ecology groups won two-thirds of the vote. In Betz, Brittany, close to the proposed nuclear site in Erdeven, ecologists won fourteen out of twenty-one seats on the municipal council. In Mulhouse, an industrial city, the ecologists increased their vote from 6 to 13 percent since the 1974 elections. Ecologists also had considerable support in large cities with an important student population. In Paris, where ecologists won an average of 10 percent of the votes, they won more than 13 percent in student neighborhoods and in those quarters with major urban renewal projects. In some middle-class suburbs of Paris the ecology vote was over 20 percent.

But electoral politics was again divisive. During the 1978 parliamentary elections, when forced to choose between left and right, the ecologists could no longer maintain their ideological independence. The tactical question of political alignment on the second ballot split the different components of the movement. Radical Marxists, equating the ecological struggle with a struggle for a self-governed socialist society, formed an electoral platform together with the PSU in the *Front Autogestionnaire* and supported all left candidates on the second ballot. The anarcho-libertarian groups around *La Gueule ouverte* argued in favor of not voting at all, reviving the May 1968 slogan, *"Elections—piège a cons"* ("Elections—trap for idiots"). Those groups already successful in the municipal elections (SOS-Environnement and Les Amis de la Terre) favored an independent strategy, arguing that the left and right were equally antiecological. The poor electoral results of the different "green lists" reflected their splits. They averaged 2.14 percent for all of France, and in Paris support dropped from 10 to 5 percent.[11] More broadly the schisms within the left that destroyed its electoral possibilities demoralized the diverse grassroots groups, leaving the antinuclear and environmental movement in a state of paralysis.

In Germany too electoral politics forced ecologists into an ideological and programmatic clarification with divisive effects. After some success in municipal elections after 1975 and a large debate in the BBU on the issue of electoral participation, ecology groups entered the *Land* elections in 1977. Only in Lower Saxony, where the continuing battle against the Gorleben reprocessing and waste disposal plant had united the ecologists, could they present a common platform. In Hamburg three ecology parties competed—a conservative green list, an environmentalized old splinter party, and a left antiauthoritarian "colored list." The colored list included the support of ecologists, feminists, prisoners rights groups, homosexuals, and a civil liberties movement formed to oppose the *Berufsverbote*. This group had the only real success. Attractive to young voters from 18 to 25 years old (25 percent of whom voted for the colored list), it gained a seat on one of Hamburg's neighborhood councils.[12] In early 1979 a group called "Alternative," formed by citizen initiatives in Berlin on the model of the colored list, won a little more than 3 percent of the vote and got some of its candidates elected to neighborhood councils. Later that year for the first time a green party won more than 5 percent in the Bremen *Land* elections, and four of its candidates entered parliament. A few weeks later ecologists had spectacular results, winning up to 15 percent of the votes in some of the communal councils and some 5 percent in the *Land* elections of Baden-Württemberg where six ecology candidates entered parliament. Here too the major support came from the youngest voters.

In Hessen ecologists split on a continuum ranging from the respectable, conservative Green Action Future, created by Herbert Gruhl, a former CDU spokesman for environmental problems, to the counterculture radical groups formed around the hero of May 1968, Daniel Cohn-Bendit. In these elections dominated by national concerns the ecologists failed to gain a significant vote. In Bavaria the splinter party Aktionsgemeinschaft Unabhängiger Deutscher (AUD) supported by Green Action Future focused entirely on an ecological program and had meager results. In Hamburg and Lower Saxony the ecologists failed to obtain the 5 percent margin necessary to be represented in regional parliaments, but matching the liberal vote, they eliminated FDP from these parliaments.

This record suggests that a coalition of citizen initiatives presenting itself less as an ecology party than an alternative way to promote post-May 1968 concerns has sufficient appeal among younger voters to be politically significant. Such a coalition represents not only a specific platform but also the young generation's alienation from the dominant political rhetoric.

The 1979 election to the European parliament confirmed these trends. Perceived as a relatively unimportant and purposeless election (only 65 percent of Germans and 61 percent of the French voted, as compared to some 85 to 90 percent electoral participation in national elections), the ecologists did rather well, with 4.3 percent of the vote in France and 3.1 percent in Germany.

Selected Targets

Organizing demonstrations at virtually every nuclear power plant site during 1976 and 1977 exhausted the ecologists. Thus the national coordinating organizations began to focus on a few selected targets: the most advanced development in nuclear technology—the fast-breeder reactor in Kalkar—and the most sensitive project in the nuclear program—the reprocessing and nuclear waste disposal facilities at La Hague and Gorleben.

Beside Windscale in England La Hague is the only operating commercial reprocessing plant in Europe at the present time. It has had operational difficulties—both minor accidents and persistent labor conflicts. As a key factor in the international expansion of the nuclear industry La Hague is at the center of the nuclear controversy.

CEA had developed a reprocessing technology for the gas-graphite technology in the mid-1950s at Marcoule. It produced plutonium for French nuclear weapons. The much larger reprocessing plant at La Hague opened in 1967 to reprocess the spent fuel from gas-graphite power plants; a reprocessing unit for pressurized water fuel was added in the early 1970s. From the beginning the health and safety of the workforce was a major problem. Reported cases of contamination increased from 280 in 1973 to 572 in 1976.[13] Most incidents did not become public. But in 1967 a severe accident took place during a site visit of three Euratom controllers. Subsequently CFDT, the union representing most of the plant's workers, asked the management to clean, repair, and modernize the center. These demands were ignored.

In 1976 the government transferred La Hague from public ownership through CEA to Cogema, a private firm. In addition to their concerns about working conditions and safety workers now feared that privatization would harm their status as public employees, leading to cuts in salary and fringe benefits. They began a strike that lasted from September 1976 to January 1977.[14]

At a large meeting organized with the help of antinuclear activists they publicized the health hazards that they encounter in the reprocessing plant. On the cover of one issue of their newsletter *Hag'Info* appeared a caricature of Hitler proclaiming: "If I were still alive, I would join the antisocial Cogema, and they would elect me as their Führer."

The strike at La Hague was the first instance of worker opposition to nuclear power. It temporarily cut an indispensable link in the production chain—the processing of material from light-water reactors to create the fuel for fast breeders, the key element of the French government's long-term nuclear program. Yet the Cogema management in Paris ignored the strike, assuming an attitude of "wait and see." Meanwhile the longer the strike went on, the harder it was to maintain unity, as hard liners from the CFDT called for persistence, while CGT activists argued for practicality.

A technological constraint finally broke the strike. Without a change of water the danger of contamination and even explosion in the cooling ponds became so great that emergency legislation forced some workers to do the necessary job. In heated discussions the workers debated the contradiction between their right to strike and their obligation to control the dangers embodied in a modern technology where certain physical and chemical processes cannot simply be stopped without public risk.

The workers failed to prevent the privatization of the firm, but they did impose a health inquiry with the participation of union officials. The committee on health and safety completed its inquiry in October 1977. Its report pointed out technical difficulties that called for stopping the expansion of the reprocessing center. It expected these difficulties to increase with the shift from gas-graphite reactor fuel (400 tons annually) to pressurized water reactor fuel; reprocessing only 230 tons of the latter would increase the radioactive emission level in the ocean above the allowed maximum per year.[15] In an appendix to the report the CFDT demanded that Cogema close the plant for at least six months to improve the safety of the working environment.[16]

Despite difficulties Cogema maintained its plans to enlarge the plant and develop it into the world's largest reprocessing facility. In 1978 the firm had signed contracts with Japan (for 2,200 tons), Sweden (650 tons), Germany (1,800 tons), the Netherlands (120 tons), and Belgium (324 tons).[17]

The French government hoped to obtain a world monopoly for reprocessing, gaining an important competitive advantage as an exporter of nuclear power plants by providing a package offer that included the use of La Hague.

Ecologists accused Cogema of taking advantage of the worldwide shortage of commercial reprocessing facilities. In several countries the licensing of new power plants is legally linked to the availability of reprocessing and waste disposal facilities; Cogema, they claimed, offered an easy way out by providing contracts that it could not really fulfill. The company, however, has sought to build up foreign confidence in the future of La Hague. When the German press questioned Cogema's capacity to fulfill its contract, the French foreign ministry refused to allow journalists to accompany a delegation of German parliamentarians on a site visit. They also prevented the parliamentarians from meeting with union activists. A French satire magazine suggested that "the French government thinks discussions between the delegation . . . and political adversaries of the reprocessing plant . . . could lead to deterioration of relations between the two countries." [18]

The ecologists, many of whom were CFDT activists, continued to improve their regional organization, mobilizing the farmers and fishermen threatened by La Hague and the nearby nuclear plant at Flamanville. During the cantonal elections of 1979 the antinuclear candidate won 45.5 percent of the vote from La Hague. A few weeks later, at the end of March, the departmental prefect sent a new DUP dossier to the municipalities near the plant. It announced that the reprocessing capacity of the existing plant would be increased to handle 400 tons a year and that a second plant with a capacity of 800 tons a year would be in operation by 1985. The CFDT and the ecologists urged the municipalities to organize local referenda to pressure public authorities and Cogema to reconsider their plans. Meanwhile the antinuclear committee in the region published an environmental impact statement and asked all municipalities to sue Cogema in the administrative courts. Two municipalities did sue, hoping that legal pressure combined with technical difficulties would increase skepticism even among the licensing authorities that have always endorsed the propositions of the nuclear industry.

In Germany, where power is more dispersed, and channels for influencing government decisions more numerous, the antinuclear movement had more impact. Even while the Kalkar demonstrations were underway, the ecologists were in court. In early 1977 the administrative court in Münster decided it could not judge the substantial issues in a case of such national importance and handed the case over to the constitutional court in Karlsruhe (see chapter 11). The reasoning of the Münster court read like the arguments of nuclear critics. The court asked if constitutional principles of legality, democracy, and division of power were still respected when such important decisions could be taken by administrators rather than by the legislature. The unsolved waste disposal problem, the environmental hazards, the management of a plutonium economy were, claimed the court, of sufficient political importance to warrant parliamentary decision.

By questioning the very principles and rules on which nuclear policy is based, the court decision reinforced skepticism among political leaders. Some SPD legislators refused to vote on the budget allocations for Kalkar until the minister of science and technology responded to a long catalogue of critical questions. Then the minister of the interior from North Rhine-Westphalia, responsible for the decision about continuing the construction, refused to sign what he considered to be an "irreversible step into a plutonium economy."

Such conflicts within the *Land* government were awkward. More than 1 billion DM had been invested in the project, and Germany had commitments toward its Dutch and Belgian partners. Thus after a long political battle the federal government assumed responsibility and submitted the Kalkar decision to the *Bundestag*. Before the vote, in early 1979, six Liberal members of the *Bundestag* announced that they would vote against the government proposal to continue construction. The CDS/CSU opposition responded with a hard-line counterproposal that not only allowed further construction but included a priori parliamentary approval for eventual operation of the plant. This created a dilemma. If the dissenters insisted on voting against the government proposal, the CDU motion would win, meaning that the *Bundestag* would approve not only construction but future operation as well. Thus, by dissenting, the six parliamentarians would have unwittingly supported a motion that was even more against their principles than the government plan. They therefore complied with the prevailing parliamentary consensus in favor of the government's proposal.

After Kalkar the nuclear controversy entered the sphere of power politics. The combined tactics of mass mobilization, legal action, electoral threat, and lobbying finally broke the political consensus. Either concern about risk or consideration of public attitudes (see chapter 8) led many politicians to question nuclear policy and shift the burden of proof for the need and safety of nuclear energy to its promoters.

Another major target of German ecologists was the proposed reprocessing and waste disposal plant at Gorleben. In 1976 Chancellor Helmut Schmidt declared in his inaugural speech that further expansion of nuclear energy could only take place if the waste disposal problem were resolved. Moreover several court decisions hinged on the development of adequate disposal techniques. Gorleben was thus of vital importance for the German nuclear industry. As the largest industrial complex in the Federal Republic its construction was both a legally mandated precondition for future licensing of nuclear plants and a symbol of technological independence and achievement. However, for antinuclear activists it was rather a symbol of the inhuman scale of technology and a key to stopping the nuclear program.

The government decided to concentrate reprocessing and waste disposal in one industrial complex, and in 1976 technicians inspected the region in Lower Saxony that was envisioned for this center. Local farmers met them in organized leisure promenades that soon turned into demonstrations. Later that year the federal government proposed three sites to the prime minister of Lower Saxony, Ernst Albrecht, and asked him to choose as soon as possible. After three months of reflection Albrecht chose a fourth site, Gorleben, in a sparsely populated region close to the East German border. Gorleben had several advantages. Local people, traditionally conservative and loyal to the government, were not expected to raise objections. Furthermore the location close to the East German border would involve bilateral discussions that only the federal government could undertake. The government of Lower Saxony could thus withdraw from responsibility, and opponents of the project would have to face the federal government.

When the Deutsche Gesellschaft für Wiederaufarbeitung von Kernbrennstoffen (DWK), a joint venture of the German utilities that builds and operates nuclear power plants, submitted its plans to the licensing authorities, organized resistance was already underway.

Ecologists asked to see the safety report submitted by DWK to the administration. At first the government refused, arguing that the report describes "a new technology that could be betrayed to Moscow, Peking, or the United States." [19] But finally it placed one copy of the thirteen-volume, 3,000 page report in an office of the ministry of social affairs. It was to be available to the public for six weeks, four hours each day. An ecologist complained: "[The prime minister] needs at least two years to examine the document. But he told us to get two or three free days to study the report in depth."

A few months later, well before the site was approved, DWK tried to buy the land. Its financial offer was very high. The normal price for one square meter in the region was 45 pf; DWK offered to pay the farmers 3.65 DM and warned them that, if they did not sell voluntarily and were later expropriated, they would receive the lower price. Few farmers resisted. But the owner of more than half of the area, Graf von Bernsdorf, did resist. He publicly declared that, although not against nuclear energy on principle, he would sell only if the best available international expertise could convince him that the project involved no major risks.

Together with the local citizen initiative Bernsdorf urged the government to form an international expert committee to assess the documents. Aware of local resistance and concerned about the forthcoming election, Prime Minister Albrecht accepted this demand. (See chapter 12.) A group of international experts critically examined the safety report and held six days of public hearings in March 1979. While they met, the Three Mile Island accident occurred. Ecologists had planned a demonstration against the project in Hannover, the capital of Lower Saxony. Stimulated by the accident, it became the largest demonstration to date (estimates of participation ranged from 30,000 to 140,000 people). The Gorleben project had to be abandoned: political as well as technical uncertainties remained.

The Gorleben International Review had another impact; it underlined the importance of scientific expertise as a political resource and indeed the leading role played by scientists throughout the nuclear controversy.

7 Science as a Political Resource

Scientists in both France and Germany played a critical role in the antinuclear movement. They have provided activists with expertise to challenge the technical reports of the nuclear establishment in public inquiries and in the courts. They have used their expertise to raise questions of safety in technical areas obscured from public knowledge. They have served as a crucial source of legitimacy and credibility as the movement sought to expand its constituency. "When a seriously dressed Dr. Engineer with a black attaché case comes to the podium in a debate, the ordinary citizen is already more impressed than he would be by someone with a beard or dirty parka." [1] Indeed scientists in the nuclear debate have been a decisive political resource, and no party could hope to gain public attention without scientific support.

The political involvement of scientists in the nuclear debate, however, has also been a source of ambivalence. While many scientists sought political involvement, they feared its implications for the image of political neutrality basic to scientific credibility. While activists sought expertise, they too feared its implications for the grassroots image of a social movement. Ambivalence, however, did not prevent important scientific involvement in the antinuclear debate; indeed it shaped the organization and the tactics of the movement in many important ways.

The Organization of Counterexpertise

In February 1975, 400 French scientists from prestigious scientific institutions (Collège de France and the University of Paris Faculté des Sciences at Orsay) signed the Appel des 400, calling on the public to reject the development of nuclear power plants. Eventually gathering some 4,000 signatures, the scientists took issue not only with the technical problems of nuclear safety but with the closed decision-making procedures that characterized this policy area. Secrecy was as central as safety in the Appel des 400:

It is troubling to see EDF elude all questions and refuse all competence outside of their official technical group. . . . It is troubling that those who promote these projects are at the same time to judge them . . . .[2]

Calling for greater consultation from the wider scientific community, this declaration shaped the subsequent organization of French counterexpertise in the nuclear debate.

After 1975 small groups of concerned scientists, most of them CFDT members working at CEA, began to provide technical advice to the antinuclear movement. As these scientists became increasingly involved in the movement, they deliberately moved beyond their scientific role, insisting that borders could not be drawn between technical and political questions.

In the United States scientists had played an important role in generating the controversy over nuclear power; many of the earliest citizen groups were organized by scientists themselves. In France, however, the antinuclear movement was well underway before scientists began to respond. While some scientists had actively opposed the development of a military nuclear force during the late 1950s and early 1960s, they had encouraged the development of a civilian nuclear program. The early siting of gas-graphite reactors on the Loire provoked essentially no response.

However, France was the home of one of the earliest European groups specifically concerned with the risks of radiation. In 1962 Jean Pignéro founded the Association contre le Danger Radiologique to investigate the dangers of radiation from all sources—X rays to atom bombs. Renamed the Association de Protection contre les Rayonnements Ionisants (APRI), this group, which included mainly secondary school teachers and physicians, began to focus on nuclear power in the 1970s but played a relatively minor role as more prestigious groups entered the nuclear debate.

Another organization of scientists, Survivre, formed in 1970 as part of an international movement for the survival of mankind. It turned its attention to the nuclear issue in 1972 after writers from its journal, *Survivre et vivre*, exposed the existence of cracks in nuclear waste storage containers at a laboratory at Saclay just outside of Paris. Survivre dissolved in 1975 as the scientist movement clustered around the Groupe de Scientifiques pour l'Information sur l'Energie Nucléaire (GSIEN), created by some of the initiators of the Appel de 400. Based in Paris, but with members in Strasbourg, Lyon, Marseille, and Grenoble, GSIEN is the most active public interest science group in France. Through its periodical, *La Gazette nucléaire* (1,700 subscribers), and a technical bulletin (250 subscribers) GSIEN provides antinuclear groups, especially Les Amis de la Terre, with information on technical and legislative aspects of nuclear power.

The Appel des 400 also generated several small groups of experts in specialized fields. Two hundred physicians signed a petition emphasizing the medical problems of nuclear energy. A group of scientists from CNRS, the Collège de France, and the National Institute for Agronomy Research formed the Geneva-based Groupe de Bellerive and Projet Alter to "study alternate energy scenarios based on the use of renewable resources." [3] The CFDT developed its own counterexpert group. Some CEA technicians, members of the CFDT, wrote the first widely circulated French book critical of nuclear energy, *L'Electronucléaire en France,* and conducted a specialized study of the problems at the reprocessing plant at La Hague. [4]

Other experts with skills and information useful to the antinuclear movement, for example, economists, have been less prone to organize. The Institut Economique et Juridique de l'Energie (IEJE), based at the University of Grenoble, published several reports questioning the nuclear choice in 1975 but has since been relatively inactive. [5] Legal advice is crucial for antinuclear activists, as they use the courts to challenge government policies. A group of lawyers created the Société Française du Droit de l'Environnement (SFDE) in Strasbourg in 1974 to deal with environmental problems including nuclear power. In its journal, *Revue juridique du droit de l'environnement,* this professional society has given wide publicity for foreign and French legal decisions pertinent to the nuclear program. The group also lobbies to improve and extend environmental legislation.

Finally, several individual scientists have played an important role in the French antinuclear movement: Lew Kowarski, a physicist and director of CEA from 1946 to 1954 (died in 1979), Alexander Grothendieck and Pierre Samuel, mathematicians, Philippe Lebreton, biologist, Louis Puiseux, economist at EDF, Marcel Froissart, physicist, are just a few of the most visible antinuclear experts. They write, lecture, and organize special studies and petitions. Others, Jean Rostand, biologist, and Jacques Cousteau, oceanographer, use their professional reputations to win public sympathy for the movement. An agronomist, René Dumont, took an explicitly political route, serving as ecological candidate in the 1974 presidential campaign.

Prior to 1975 few German scientists were engaged in the antinuclear movement. As in France an early movement of concerned scientists formed in the late 1950s when research facilities were first developed at the large nuclear centers at Karlsruhe and Jülich. At that time some well-known scientists feared that the complex research structure financed by both state and private capital was in fact intended to enhance military capacity. Eighteen distinguished physicists, including Otto Hahn, Werner Heisenberg, and Karl Friedrich von Weizsäcker, expressed this fear in the

1957 Declaration of Göttingen.[6] But they focused on the military use of atomic energy, and scientists with few exceptions looked favorably on the peaceful uses of atomic energy. When the *Bundestag* organized a series of regional study groups in 1973 to study the environmental impact of nuclear energy, few scientists participated. Even later at Wyhl, of the fifty-one scientists who testified at public hearings only nine were critical of nuclear power. The expansion of the German nuclear program in 1974, however, did provoke a response.

As German scientists organized to develop an independent source of expertise, they formed no centralized association equivalent to GSIEN. Instead small working groups were scattered in universities throughout Germany.

In the 1960s groups of scientists with sympathies toward the political left formed in Münster, Marburg, Bochum, and Tübingen with names such as the League of Democratic Scientists, Scientists and Politics Work Group, Life Protection Work Group. They sought to reorient scientific efforts and promote research in the service of nonestablishment interests. Since the mid-1970s these groups of critical scientists have worked on environmental and antinuclear issues. However, many scientific petitions also endorsed nuclear power, a fact used by officials to prove that nuclear power is accepted by the rational elements of society.[7]

The University of Bremen, established at the end of the 1960s to promote research directed to contemporary social problems, is the center for the most politically active antinuclear scientists. In 1973 some scientists at the university organized a program called Science at the Service of the Underprivileged. This program published a book, *66 Ways to Understand the Nuclear Industry,* that urged the scientific community to appraise critically the nuclear program.[8] The Bremen movement grew after 1975: students wrote antinuclear dissertations, the staff organized a seminar on the nuclear controversy, and professors publicly criticized the nuclear plant licensing process. A group of young physicists wrote a critical book, *Nuclear Waste Disposal . . . the End of an Expensive Dream,* presenting the problem of nuclear waste as a moral and political issue.[9] Bremen scientists also extended their actions outside the university. A nuclear physicist, Jens Scheer, was forced from his teaching post, and fifteen activists from Bremen University served jail sentences for participating in the Brokdorf demonstrations.

Another center of outspoken activity is the University of Heidelberg. Working groups at the university translated articles from American journals and wrote their own studies of the impact of nuclear energy, including a critical analysis of official reports monitoring radioactive pollution at power plant sites.[10] An interdisciplinary team of twenty-seven biologists, chemists, physicists, physicians, and veterinarians, the Tutorium Umweltschutz, criticized the official radiological expertise presented to the administrative court at Wyhl and introduced its own report as counterevidence. The university administration took disciplinary steps against this group which had used the university name in their publications without authorization. In 1977 the scientists created a private institute outside the university.

The importance of legal actions in the German nuclear debate helped to shape the organization of expertise. Scientific evidence was clearly influential in the growing number of court cases, and German ecologists sought to organize information in a form usable in legal proceedings. With this in mind in 1977 Siegfried de Witt, the lawyer who defended the antinuclear case at Wyhl, helped to create the Öko-Institute in Freiburg. This Institute employs five full-time scientists and ten others on a part-time basis. It holds conferences, publishes a newsletter and brochures, and maintains contact with citizen initiatives in Germany and in other countries. Its main purpose is to link scientific expertise and legal advice and to provide citizen initiatives with useful technical information on environmental problems. In 1979 the institute opened an office in Hannover, close to the Gorleben site.

The different organization of counterexpertise in Germany and France reflects differences in their research systems, their general political climate, and the traditions of their scientists. Most German counterexperts come from universities and their groups are scattered throughout the German academic system. In France activist groups are relatively centralized, mostly located in Paris, and they include not only academics but research personnel from government agencies and firms, in particular CEA.

Because of the public funding of R & D in universities and mission-oriented "big science" laboratories, counterexperts in both countries are in the peculiar position of working in organizations financed by the very governments they are challenging. In general universities are more autonomous than government research centers. Indeed in Germany nuclear critics are numerous in universities, but few critical voices come out of the research centers in Karlsruhe or Jülich. In France the initiative to form GSIEN came from scientists working in academic institutions (the Collège de France and the Faculté de Sciences in Paris), but scientists working in the very centers of the nuclear enterprise (CEA and EDF) have also been publicly critical of nuclear power.

While the institutional affiliations of counterexperts vary, the scientific disciplines they represent are often similar. In both countries biologists were especially sympathetic to the antinuclear movement; as an environmental controversy it gave them ample opportunity to promote ecology as a scientific field. While some physicists, especially in France, were very active, others, especially those working close to the engineering professions, were not sympathetic and were turned off by the general criticism of technology pervading the antinuclear literature.

A number of professional groups had pragmatic reasons for participation in the antinuclear movement. Lawyers clearly welcomed the possibilities of professional expansion in this area: this was explicit in the charter of the French professional association of environmental law. In Germany sociologists and political scientists who had been active in peace research or conflict analysis during the 1960s now saw a possibility to turn their attention to the citizen initiative movement as a harbinger of social change. Similarly for French sociologists long concerned with social movements the controversy provided a new area of investigation. But in both countries social scientists were initially skeptical of ecological themes, entering the debate only after the movement had developed an explicit political analysis. In addition their arguments were often dismissed as unscientific or irrelevant by hard scientists. Ironically in the adversarial climate of the nuclear debate physicists and engineers often focus their arguments on economic issues, while social scientists talk of technical risks. This often ends in reciprocal accusations of incompetence, tempered only by mutual interests in preserving the credibility of expertise.

A further structural factor has shaped the organization of counterexpertise. Most active participants are either young or old in terms of their scientific career. Attacks against the dominant scientific opinion, which in this case favors nuclear energy, are most often formulated either by powerful scientists who take no risks by presenting a critical position or by young scientists who were socialized during the student movement of the 1960s. This group, tending to translate problems in political terms, is by far the dominant voice among the antinuclear experts. For many of them the adversarial climate of the nuclear controversy provided an opportunity for public acclaim and influence helpful in their professional careers.[11]

Scientific arguments in such disputes cannot be dissociated from political and career considerations. The positions assumed by scientists with respect to nuclear power can be explained in structural terms. Those closest to the center of political and economic power tend to formulate pronuclear arguments, while those generally critical of the existing social and economic order tend rather to oppose nuclear power. An important source of critical views has been the growing class of research personnel and technicians in subordinate and increasingly insecure positions. Also many of the antinuclear scientists come from fields such as medicine and the social sciences peripheral to the "center" of nuclear science and engineering. They entered the debate at a late stage when the limits of a narrowly defined technical analysis were apparent. Their participation helped to broaden the dimensions of the debate and also coincided with their own interest in professional expansion. In sum scientific participation on both sides of the controversy reflected attitudes toward the established social order both within the scientific enterprise and in the society at large.

It is not surprising that at first the social characteristics of counterexperts led observers to disparage their seriousness, credibility, and respectability; low status in the scientific community was presented as proof of lack of competence.[12] Later the involvement of some prestigious older scientists and Nobel prize winners in the antinuclear debate tempered these attacks. But at the same time the willingness of scientists to enter the debate created tension within the scientific community, characteristically ambivalent about the wisdom of political participation.

The Ambivalent Tactics of Scientific Activism

Scientists engaged in nuclear debate with reluctance. Once involved, their tactics revealed continued ambivalence, especially as the discourse turned to increasingly political themes. As they entered the nuclear debate, most scientists expected to play a rather limited technical role. Indeed scientists at first focused mainly on technical questions—the design of reactors, the problems of waste disposal, and the potential environmental and health impacts of specific siting plans. But the role of critic very quickly engaged the scientists in public dispute. One of the early disputes in France took place in June 1974 over a report from a government-sponsored laboratory predicting that the environmental impact of the nuclear program would be negligible. Some members of the laboratory's science council took issue with the report, accusing its authors of servility to the government. They initiated a counterstudy on the environmental consequences of nuclear power.[13] In Germany dissenting scientists first aired their disagreements in public in November 1975 in a series of open letters about nuclear energy.[14]

Some of the early public disputes concerned the health effects of nuclear power. A French group from Saclay published medical dossiers suggesting that industrial hazards are far worse than reported. Workers, these scientists claimed, were poorly informed and afraid to complain, and industrial secrecy surrounded the issue of worker safety as laboratories tried to avoid regulations that would delay production.[15] In Germany the Lower Saxony medical journal published a series of papers indicating the disagreement about radiation risks among physicians.[16] A biology research team from Heidelberg wrote a detailed report on the radiological impact of the Wyhl nuclear plant, taking issue with official data.[17] A dispute focused on the appropriate use of the unit "curie" instead of "rem" as a measure of radiation, critical scientists arguing that the definition of rem depends on an arbitrary coefficient.[18]

Technical disputes also focused on economic questions. Economists from the IEJE in Grenoble took issue with the economic reasoning that backed the nuclear program. "Too many uncertainties concerning comparative costs remain to consider nuclear power the major energy source for the next thirty years." They insisted that social costs—the costs of environmental protection, of social security for the illnesses of nuclear workers, of preventing sabotage—be incorporated into the economics of nuclear power. They questioned the methods of predicting future levels of energy consumption on the basis of past extrapolations. German economists attacked the prevailing wisdom that more energy means more jobs, arguing that energy-intensive technologies were more likely to destroy than create jobs.[19] This argument became central to discussions at the 1977 SPD workshop on Energy, Employment, and the Quality of Life.

That experts disagree is hardly surprising, but the public airing of these disputes has engaged scientists in an increasingly political dialogue. In France, with comparatively few ways to directly influence policy, scientists focus their efforts on swaying public opinion. To catch the public eye, technical critiques appear with flashy and often macabre covers. Technical facts, graphs, and equations are interspersed with irony and metaphor. The APRI journal proposed a new calendar that begins in 1942 when Fermi's Chicago pile became critical.[20] Louis Puiseux's data-filled critique calls the reactor a daughter of the bomb.[21]

French scientists also engage in global political analysis, identifying nuclear power less as a technology than a choice of civilization.[22] This perspective emerged in the mid-1970s in a few marginal magazines such as *ImpaScience* devoted to critical analysis of science. But prior to 1977 most French scientists avoided political and social analysis. GSIEN's first book, *Electronucléaire: danger*, contained only six pages on proliferation and other sociopolitical issues associated with nuclear energy. By 1977, however, scientists began to focus as much on the characteristics of a nuclear society as on the technical questions of risk. By 1977 *La Gazette nucléaire* was systematically addressing broad social issues. A group of French scientists, commenting on the political problems raised by nuclear power, saw the antinuclear movement as a new political force, "an expression of the common will for radical change in the society." [23]

The approach of German scientists, often involved as experts in court actions, has been more pragmatic, but even here the sociopolitical implications of nuclear power entered their dialogue. As the scientists' discourse turned to political issues, it often assumed an apocalyptic style. Some expressed their concern about a police state. Bodo Manstein, biologist and environmentalist, writes: "Protection against external threats to nuclear facilities must be established, for signs of civil war can already be discerned; a police state is inevitable." [24] A group called Aktion pro Vita published a leaflet listing the precautions necessary in case of a nuclear accident. [25] Popularization flourished. A book by several German scientists, *Atomic Dilemma*, seeks to clarify the connections between the social, political, military, and industrial interests involved in the nuclear choice. Holger Strohm, an ecology activist with a physics background, wrote a best-seller called *Friedlich in die Katastrophe*. [26] Robert Jungk's *Atomstaat* was for several months on top of the German best-seller list.

The scientific discourse paralleled the changing tactics of the antinuclear movement. After 1977 the scientists' organizations paused to reconsider their efforts, turning increasingly to study and promote alternative energy sources. *La Gazette nucléaire* devoted three issues in 1978 to geothermal, solar, and biomass energy sources. A special issue of the largest French consumer magazine, *Que choisir*, focused on alternate energy. [27] The Öko-Institute called its first publication *Alternate Energies: Other Ways of Thinking and Believing*.

While clearly an active and critical force within the antinuclear movement, scientists remain ambivalent. Even when writing in a political mode, they emphasize their technical qualifications, their titles, and degrees, and they sprinkle their articles with formidable equations and footnotes to establish the solidity of their claims. But the political dynamics of an active social movement inevitably encourage careless comparisons and violations of professional standards that bring sanctions from colleagues. Many scientists, socialized to see their work as neutral and expecting to air their disputes within the closed community of science, are appalled at the public and political nature of the disputes and fear the consequences for the autonomy of science and their ability to maintain control over research.

While facing collegial pressure, activists also face material constraints. The threat of losing employment is greater in Germany than in France. German nuclear industries urge their employees to avoid taking critical public positions that might affect the company's interests. The Technischer Überwachungsverein (TÜV-Hamburg), responsible for technical control of power plant operation, fired an engineer for taking part in the Brokdorf protest, and in a number of cases internal regulations originally intended to protect patented technical information were extended to prevent employees from offering controversial opinions on technical issues.[28] Moreover the fact that many scientists in universities and research centers are civil servants has different implications in Germany than in France. German civil service candidates must submit to routine investigations of their political past before they are granted tenure. The *Radikalenerlass* (see chapter 12) has tempered political fervor and favored pragmatic behavior. Scientists can be expelled from their jobs for challenging the constitutional order. In addition, to avoid conflict with university administrators, German antinuclear academics sometimes locate their controversial research in private institutes specifically created for that purpose. These, however, are vulnerable. Dependent almost entirely on government funding, their existence rests on timely political conditions. In areas of public controversy, the German government often finances such institutes to prove its concern with unbiased expertise. As the political agenda shifts, funding can disappear.

In France the job security in civil positions allows outspoken radical political analysis. Furthermore most French scientists belong to a union. Even when officially pronuclear, the French unions afford protection to critical government employees. Thus CEA employees who signed the Appel des 400 received only a letter of rebuke from the CEA administration. A nuclear physics group within CNRS, the government-sponsored research institution, took a critical position against the nuclear program and was able to publish its conclusions.[29] Even in France, however, activists feel professional pressure: an engineer from Fessenheim who expressed concern about the materials used in the piping was transferred to another region.[30] Some government employees who took part in a study organized by Michel Bosquet refused "for professional reasons" to let their names be used.[31] Thus career pressures combined with collegial considerations to create a certain self-censorship among scientists that tempered their political involvement.

The ambivalence of scientists as they engage in the nuclear debate reflects a larger issue. Possessing knowledge that is indispensable for politics and public policy, scientists have been increasingly aware of the changing dimensions of the power and responsibility embodied in their professional position.

Social Responsibility of Science and the Distribution of Expertise

Intellectuals in France have a long history of participation in public affairs. Expected to articulate the problems of society, the French intellectual community has often served as a powerful pressure group. During the May 1968 strikes and the wars in Indochina and Algeria they were outspoken figures, committed to represent the critical consciousness of the nation. When French scientists entered the nuclear debate, they copied the style of their predecessors. As intellectuals, imbued with a deeply held belief in the power of ideas, they were more inclined to take a global view of the nuclear issue than to dwell on narrow technical issues.

In Germany, where pragmatic technical arguments carry considerable force, especially in court actions, activists felt a responsibility to maintain a technical discourse. They perceived their role less as intellectuals responsible for articulating general problems of society than as a technically competent group responsible for redressing inequities in the distribution of expertise.

Efforts to define their social responsibility create basic dilemmas for scientists. Proestablishment figures such as the physicist Louis Le Prince-Ringuet and many of his colleagues have defended the nuclear program as part of their responsibility for the future well-being of the nation. France, they feel, must move to the nuclear age: "I do not think that we should wait for the public to be fully informed or for all risks to be completely overcome before committing ourselves to nuclear energy." [32] A group of pronuclear engineers from Saclay accused their antinuclear colleagues of irresponsible appeals to public fear when they should rather be working to increase safety. Antinuclear scientists, however, are convinced of their collective responsibility to call public attention to technological risks.[33] "One can always hope that some hypothetical advance in technology will bring all solutions, but this is an attitude of faith, not of science." [34] Moreover many activists also feel their responsibility extends to involvement in political issues because, claims a GSIEN petition, "the choice labeled scientific is in fact a political choice." [35]

Defining the issue in political terms, scientists thought of their social responsibility in terms of distributing technical knowledge. They attacked the closed character of the nuclear enterprise. A CNRS group complained that they had to rely on foreign sources of information about the risks of nuclear power because of their own government's secrecy. The lack of public information on procedures in case of an accident became a major issue stimulated by the discovery in Germany of some secret evacuation plans. Similarly, when eighty researchers and technicians from the physics institute at Lyon appeared at the inquiry on the fast-breeder reactor at Creys-Malville, they concerned themselves less with the technology than with the inadequacy of the information available to the public about the breeder reactor.

The scientists' efforts to affect the distribution of knowledge, however, contain several contradictions. When scientists stress the importance of a more equitable distribution of expertise, they assume that this will help better to inform the public debate and democratize the decision-making process. Yet their efforts simply transfer public dependence to a different source of expertise. When scientists expose the conflicting technical views within the scientific community, they raise public doubts about the neutrality and independence of science; yet by engaging in a debate, they too seek credibility as a source of expertise. When scientists use their expertise to bring legitimacy to the nuclear opposition, they are working with a movement partly based on mistrust of precisely the expertise that they themselves represent.

The importance of scientists for the antinuclear movement, however, should not be underestimated. Their expertise on technological risk provides an indispensable source of credibility at a time when science is increasingly the dominant ideological discourse. But what is the significance of this trend? The voices of critical scientists combined with public pressure force governments to extend controversies to the public arena, changing the rules of technical debate. Criteria of fairness and equal representation replace those of scientific authority. Those scientists who participate in public forums are selected to represent both pros and cons. This prolongs debate, and it also creates opportunities for scientists in bureaucratic functions in an increasing number of policy areas. Conflict is said to indicate the declining power of the scientific enterprise—to reflect a crisis of science and even a crisis of rationality. However, the material side effects of scientific and technological debates, measured in the amount of resources put into new research and the creation of new positions, rather indicate that scientists have an interest in maintaining a degree of publicly expressed dispute. But, while such disputes give visibility and influence to critical scientists, they often mask the institutional fragility of their enterprise. For under politically less favorable conditions—without the backing of a social movement—their voices can easily be dismissed.

III THE NUCLEAR OPPOSITION AS A SOCIAL MOVEMENT

8 The Social Base of the Antinuclear Movement

Violent events often mirror the society that produces them.[1] Indeed social movements have historically emerged as a reflection of significant and disruptive changes in economic and social structure.[2] These changes often produce new social groups that seek to control the conditions of their existence through conflict with the established order.[3]

In both France and Germany the antinuclear movement followed a period of economic change, affecting especially the agricultural community and the educated middle class. It took place at a time when the postwar baby boom generation entered the labor market and political life. Both countries experienced a decline of the agricultural population, a rapid and often disruptive pattern of urbanization, and the rise of a new educated and skilled social class with high expectations but relatively little political clout. The values accompanying such structural changes placed increasing demands on the political system; yet the outlets for such demands—the parties, the parliaments, and the unions—were less and less able to serve their traditional role as channels for political mediation. In such circumstances protest movements tend to form outside of existing political structures; and the antinuclear movement is a case in point. This chapter suggests some general social and economic features that contributed to concern about nuclear power in France and Germany and then describes the patterns of political conflict in each country that shaped the way antinuclear attitudes became transformed into political action.

Social and Economic Change

Several socioeconomic trends in France and Germany have converged to create a receptive market for ecological ideals. Many of these trends apply as well to other industrialized nations, and indeed to France and Germany during other periods of history. Here their convergence is important for understanding the diverse constituency of the antinuclear movement and explaining the complex thematic content of its discourse.

The movement has found a base of support from certain traditional groups displaced by the rapid social and economic changes after World War II. To them nuclear power represents a further threat. For another constituency the antinuclear movement represents their growing concern with the quality of life, a struggle for clean air, water, and space. The major support for the movement, however, comes from the young—from an educated youth culture seeking to translate the values expressed during the student movement of the 1960s into a political discourse.

France and Germany experienced different patterns of economic growth until about 1960. Between 1950 and 1960 the growth index in the Federal Republic of Germany was 8.5 percent, almost double that of France (4.8 percent). The modern development of the industrial economy of France, delayed by the internal instability of the Fourth Republic and the colonial wars, only took off after the late 1950s. But by the mid-1960s France's growth rate (5.8 percent) exceeded that of Germany (4.8 percent), and the patterns of economic development have subsequently converged.

Economic changes have had important implications for employment. For France, given the long stability of her economic structure, the changes that followed the postwar expansion were especially profound. The agricultural sector, occupying around 40 percent of the working population until 1946, dramatically declined to 10 percent by 1975, while the service sector increased from 34 to 52 percent in the same period. The proportion of workers in independent professions and management positions increased between 1954 and 1974 from 8.7 to 19.7 percent, but significantly by 1977, 45,000 people in this promising professional sector were unable to find jobs as compared to 14,000 in 1971.[4]

In Germany changes that began before World War II accelerated. The agricultural population, already down to 22 percent by 1950, further declined to 7 percent in 1975. The rate of employment within the manufacturing sector has hardly changed in Germany since the beginning of the century, but as in France the increase in the tertiary sector has been continuous and significant—from 33 percent in 1950 to 47 percent in 1975—with particularly marked increases in professional and managerial work.[5]

The necessities of postwar reconstruction at first precluded the social conflicts that often result from such rapid economic growth. For many growing prosperity and consumerism reduced the tensions from occupational change. However, industrialization also influenced patterns of migration and urbanization, affecting daily life in ways that have fostered a growing consciousness of environmental values. Before 1960 more than half the French population still lived in rural areas. By 1973 over 73 percent of the people lived in cities with over 50,000 inhabitants. Industrialization brought many people directly from the country to the largest cities, mainly Paris.[6] Just as industrialization turned farmers into urban workers, so cityfolk became weekend farmers. About 250,000 French people owned second homes in 1946, 1,683,000 in 1975, and more than 1,800,000 in 1978.[7] In Germany urbanization began much earlier, and the concern with environmental values can be associated rather with the suburbanization that began in the mid-1960s.[8]

For city dwellers, seeking relief in the country or the suburbs, the encroaching problems of expanding industrialization became an almost personal insult. These changes had a profound effect on attitudes toward the physical environment and created a constituency receptive to the values expressed in antinuclear and ecology discourse.

But demographic and educational trends appear to be the most important factor in the formation and the activism of the movement. Between 1950 and 1960 the population grew from 50.2 million to 61.5 million people in the Federal Republic and from 41.6 million to 50.8 million in France. The generation of the postwar baby boom was accommodated by significant expansion of the system of higher education. In Germany attendance at universities more than doubled between 1965 and 1975; in France the number of students tripled during this period. These education systems began to mass-produce a new class—an educated and critical sector imbued with high expectations concerning their social environment, independence and autonomy, and future political role. But as this large generation entered public life, the opportunities were not there. Nor was their influence commensurate with their political expectations. Socialized to question authority, they found themselves dependent upon it. Unwilling to give up their values, some turned to alternative cultural styles, others to radicalism and criticism of political and economic trends. The activism of this period was short-lived, but the pervasive social influence of this youth culture laid the ideological ground for the ecology movement of the 1970s, with its concern about authority and populist tinge.

Antinuclear Attitudes

Socioeconomic changes in both France and Germany had greatest impact on the peasantry, on the traditional factions of the petite bourgeoisie, and on the so-called new middle class, concentrated in the younger age groups with higher educational training. Although their agendas are very different, these groups form the basis of the environmental and antinuclear movements. The peasantry, especially in France where changes have been

more recent than in Germany, seeks to defend its way of life. They antici-
pate that the construction of nuclear plants will cause expropriation of
their land and environmental damage to their crops and livestock. The
peasantry's defensive reaction against further industrialization is reflected
less in its general attitude toward nuclear power than in its radical behavior in siting disputes.

The traditional factions of the petite bourgeoisie in small manufacturing firms and commerce oscillate in their attitude toward nuclear
power. Seeking to survive and maintain economic independence in the
context of continued industrialization and concentration, they sometimes
perceive nuclear power negatively as a product of large industry and sometimes positively as a guarantee against economic crisis.

The attitudes of the expanding modern middle class reflect its concerns
about a better quality of life. Many antinuclear activists are students.
Others are intellectuals who were socialized during the student movement
of the 1960s with a concern about the social and political implications of
technological change and a skepticism of authority. This group fully expected that its knowledge and political skills would yield political
influence. Its values and expectations have thus brought this sector into
conflict with the older administrative elite committed to the ideology of
growth. They have also attracted many students who constitute an important constituency for the antinuclear movement. Enjoying material security and stable income, this constituency represents what some have called
a post-material culture,[9] and others a new class.[10] Its concerns about the
quality of life and the preservation of the environment reflect values that
derive more from lifestyle goals than from basic economic needs. Its skepticism about authority brings to the movement a critical perspective on the
political status quo.

Many professionals from education and the media are devoted environmental activists. Producers and diffusers of cultural goods, they are in key
positions to disseminate the values and amplify the goals of the movement.
As they transformed their own values into ideological trends, they have
helped to shape the public attitudes that support the antinuclear
movement.

According to opinion polls, after 1974 the increase in opposition has
been considerable in both France and Germany. In France the percentage
of the adult population in favor of this technology declined from 74 percent in 1974 to 47 percent in 1978, and those opposed increased from 17 to
42 percent. In Germany those in favor declined from 60 percent in 1975

to 53 percent in 1977; those against increased from 16 to 43 percent.[11]

Surveys in both countries (tables 8.1 and 8.2) suggest the broad range of groups concerned about nuclear safety. Younger people and those most highly educated are the most concerned about nuclear power. In France women are clearly more hostile toward nuclear energy than men. In Germany women have a high rate of no response, and only one-third of respondents perceive no threat, indicating a greater skepticism toward that technology than among men.

Significant fears prevail among blue-collar workers, despite the pronuclear tendencies of the unions (see chapter 4). A high percentage of blue-collar workers are against nuclear power in France, and unskilled blue-collar workers in Germany are significantly more concerned about safety than any other group. While German farmers have been very active in specific nuclear controversies, relatively few perceive nuclear power as a threat. In France, on the other hand, where farmers have experienced more recent disruption from economic and technological change, their level of opposition is greater.

Most in favor of nuclear energy are white-collar employees, businessmen, and the professions, although interviews suggest differences between the independent professions, such as law and medicine, and the salaried professions, such as teaching, writing, or research, the latter being far more critical of nuclear power.[12] German surveys disaggregate the responses according to organizational affiliation, disclosing some interesting results: in the larger production units workers are less inclined to perceive nuclear power as a threat to safety, suggesting that people integrated into the production process, especially in the modern highly organized sectors, tend to have relatively high confidence in technology, including nuclear power. Nonunion members are more concerned about safety than union members, but the difference is not as large as one might expect given the unions' official support for nuclear power. Less than half of the union members responding to the German study accepted the official union approval of nuclear energy. In a similar manner 57 percent of those stating a preference for the pronuclear Communist party in France claimed to be against nuclear power.[13]

Protestants and those without religious affiliation in Germany perceive somewhat more of a threat from nuclear power than Catholics. Protestant clergies have on several occasions officially supported the antinuclear movement, perhaps creating more awareness of the issue within their congregations.

Table 8.1
Attitudes toward nuclear power: France

Question:
Are you for or against the development of nuclear power plants?

	Against	For	No opinion
Total population (1,000)	42%	47%	11%
Sex			
Men	39%	54%	7%
Women	46%	42%	12%
Age			
18–24 years	54%	38%	8%
25–34 years	55%	38%	7%
35–49 years	37%	49%	14%
50–64 years	33%	55%	12%
65 years and above	37%	54%	9%
Occupation			
Farmers	42%	41%	17%
Small business	44%	43%	13%
Professionals and big business	29%	64%	7%
White collar	44%	49%	7%
Blue collar	48%	39%	13%
Retired, nonworking	40%	53%	7%
Education[a]			
Primary	33.5%	51.0%	15.5%
Primary-superior	32.0%	62.0%	6.0%
Secondary	33.0%	62.0%	5.0%
Technical-commercial	30.0%	63.0%	7.0%
Higher education	40.0%	55.5%	4.5%

Source: SOFRES; reported in *Figaro,* December 1978.
[a] This item is from 1976 survey, asking the same question (SOFRES, 1976; reported in Francis Fagnani, *Nucleopolis*).

Analysis of the news media from various constituencies suggests that in Germany people at the bottom of the social hierarchy focus their concern about nuclear power on questions of risk, while those more highly educated are sympathetic to the political arguments of the antinuclear movement. In France the fear of risk is an important argument, but sociopolitical considerations pervade the debate. There generally appears to be greater support for nuclear power in France than in Germany, but in both countries an important proportion of the population—around 40 percent—express skepticism. A much larger proportion of those surveyed in Germany at nearly every level chose to express no opinion at all.[14]

The opposition to nuclear power has brought together a heterogeneous mix of interests, cutting across several social classes with very diverse agendas. However, the broad base of social support for the antinuclear movement does not necessarily translate into active political participation. A great many more people will sign petitions than march in demonstrations, and sympathies expressed in attitude surveys are seldom reflected in votes. In one French survey 72 percent of those interviewed declared that a candidate's position on nuclear energy would have no influence on their vote.[15]

The vanguard of the movement, willing to act on its ideological concerns, is a young, highly educated group, often in the precarious and frustrating socioeconomic status characteristic of people in their student and early post-graduate years. For others direct political activism is based more on cold assessment of the possibilities of real influence than on their vague ideals. This assessment is influenced by traditional patterns of protest and the anticipated consequences of political action.

Patterns of Political Conflict

France is often presented as an arena of social conflict and flamboyant demonstration. Germany, in contrast, tends to be stereotyped as a nation of discipline, harmony, and obedience. These stereotypes are no more true than any other; the image of harmony, for example, hardly corresponds to the turbulent reality of German history. But to some extent the picture does help us understand the present relationships among major political and professional or representative organizations, relationships marked by persistent conflict in France and the search for compromise and consensus in Germany that has followed its divisive history. Cultural attitudes toward conflict and consensus are also reflected in the interaction between

Table 8.2
Attitudes toward nuclear power: West Germany

Question:
Do nuclear power plants represent a threat to the safety of the population, or is there no reason to be concerned about safety problems?

	A threat	No threat	No opinion
Total population (1,196)	41%	37%	22%
Sex			
Men	41%	42%	17%
Women	42%	33%	25%
Age			
18–24 years	55%	28%	17%
25–34 years	44%	40%	16%
35–49 years	37%	43%	20%
50–64 years	41%	38%	21%
65 years and above	37%	32%	31%
Occupation			
Farmers	28%	72%	0%
Professionals	19%	51%	30%
Civil service, employees (white collar)	40%	42%	18%
Skilled blue collar	42%	45%	13%
Unskilled blue collar	56%	29%	15%
Education			
Primary	43%	32%	25%
Primary and apprenticeship	39%	39%	22%
Primary and professional	41%	42%	17%
High school and university	47%	37%	16%
Production unit			
−100	41%	36%	23%
101–2000	44%	38%	18%
2000 +	35%	47%	18%
Union affiliation			
Labor union members	40%	43%	17%
Nonunion members	42%	36%	22%
Religion			
Catholic	37%	37%	26%
Protestant	44%	38%	18%
None or other	54%	35%	11%

Source: Poll INFAS, May 1977.

these two governments and their citizenry and in the traditional ways of dealing with controversial political issues such as nuclear power.

In France belief in the need for a strong and highly organized central authority capable of containing the centrifugal forces within the society has fostered a social system often described in terms of *incivisme,* or lack of civic spirit. Sporadic confrontation, demonstration, and rebellion have marked the relationship between the citizen and the state in France: the food riots of the eighteenth century, the silk workers' rebellion of 1834, the tax rebellions of the 1840s, the wine growers' protests of the early twentieth century, the demonstrations over the colonial question in the 1950s, and over regional autonomy in the 1960s and 1970s. Historically, argues Alain Touraine, such incidents of collective violence were not signs of social breakdown but rather constituent elements of the French political process.[16] Given the centralization of political authority in France, explosive rebellions over specific issues invariably translate into criticism of Paris-based policy. Open conflict critical of state policy is accepted as part of everyday life. The citizen in France, claims Stanley Hoffman, derives moral and political identity only in opposition to the state.[17] Moreover with few mechanisms for conciliation in areas of disagreement citizens have little alternative but dramatic protest when they feel their rights are compromised.

This political style has characterized France's labor disputes. Unions in France are hardly recognized by government as valid social partners in the decision-making process. Refusing to be integrated within the system, they are rather organized as a counterpower; strikes are not only a means to obtain wage increases but also a way to demonstrate labor's capacity for mass mobilization. Contracts between employers and unions are not binding, and wildcat strikes occur frequently as an accepted instrument of conflict.[18]

Foreign policy is another important cause for social conflict in France. The street demonstrations over these issues were widespread as the left confronted the right over the question of Algerian independence. Most such demonstrations are organized or endorsed by a strong political apparatus that guarantees order. In this climate of permanent, but supervised, social conflict, acts of symbolic violence against property seldom turn to actual violent confrontation.

The existence of a Communist party that in principle opposed the dominant social order has in the past encouraged the translation of conflict in French society into the logic of a right-left polarization. But the student movement of the 1960s, occurring just when the Gaullist regime was recovering from the right-wing reaction to Algerian independence, broke the normal pattern of cooperation with established political ideologies. Immersed in anarcho-libertarian thought and a Freudian interpretation of Marxism, and strongly influenced by the writings of Wilhelm Reich, Henri Lefebvre, and Herbert Marcuse, the students vigorously attacked the large organizations of the working class, in particular the Communist party, for growing cultural conservatism. They identified marginal and minority groups as the main revolutionary force.

The student movement failed to translate its objectives into concrete political terms in May 1968 and disappeared about as quickly as it had invaded the political scene. However, the sympathetic waves of strikes throughout the country in May and June 1968 suggested that the students had addressed some widely shared hopes and frustrations, especially among the younger workers. The momentary solidarity during those extraordinary months created expectations that later found expression in a host of new social movements organized around such issues as regional autonomy, women's liberation, gay rights, protection of migrant workers, ecology, and nuclear power. In addition the organizations of the extreme left (the Gauchists) were strengthened after 1968. These groups adopted the tactics of the student movement, seeking to mobilize the public to create pressure on political organizations. They reinforced each other, constituting a diffuse social network that could be mobilized whenever interests were shared; the cooperation between antinuclear, environmental, and regional autonomy groups became a case in point. Despite some efforts by the new Socialist party to integrate these movements into a political coalition, they have maintained their independence from the traditional political scene.

In Germany the experiences of the Weimar Republic and the history of fascism united the political leadership in a strong desire to avoid conflict. After the war reconstruction had required the collaboration of capital and labor in a social partnership *(Sozialpartnerschaft)*. Subsequently the German labor unions and industries have participated as partners in an institutional bargaining process. Unions are less a counterforce than an institution for developing compromises, and strikes are less a weapon than a bargaining instrument in a contractual relationship between unions and employer organizations. Strikes are much less frequent than in France, causing on the average less than one-third the number of person-days of work to be lost each year.[19]

Harmony has been reinforced by the relative stability of the German political party system, in part maintained by the official anticommunist ideology that has persisted since the creation of the two German states in 1949.

In this climate of compromise among the power elite, social and political conflict tends to be expressed outside of established political channels and presented in moral rather than political terms. For example, in the 1950s extraparliamentary groups formed to oppose nuclear rearmament and military integration into NATO. Reminiscent of the antimilitarist movement of the 1920s, the arguments against rearmament referred to the war and to Germany's moral guilt. Radical democratic attitudes and a moral commitment to prevent a Nazi revival shaped this movement. Most of its leaders came from a religious, often Protestant, background, interpreting their Christian engagement as a commitment to democratic rights, freedom, and peace.

In 1957 to 1958, when the German government proposed to equip the army with nuclear weapons and to place U.S. nuclear troops on German territory, the opposition party SPD and the major union DGB spoke out strongly against nuclear armament, even raising the possibility of a general strike. SPD and DGB officials became leading figures in the Committee Kampf dem Atomtod, created in 1958. The churches and the critical intelligentsia helped to mobilize an antiweapons campaign, and citizen initiatives formed throughout the country. An opinion poll in 1958 showed an overwhelming majority of 83 percent against nuclear armanent and 92 percent in favor of a general strike to prevent it.[20]

When the government dropped the German nuclear armament plan, the antiweapons movement turned to the issue of international nuclear tests. Copying the British Easter marches against nuclear weapons, German radical democratic pacifists organized massive Easter marches in the early 1960s, involving 100,000 people.[21] But SPD and DGB withdrew from the actions, reluctant to collaborate with social movements that they could not control. By then the SPD had accommodated its policy to the reality of Adenauer's government, supporting western integration, rearmament, and a free market economy. As this became apparent, the campaign for disarmament widened its critical political analysis and began to mobilize an extraparliamentary movement around social issues they felt were neglected by the major parties.[22]

This experience in extraparliamentary organization was significant for later relations between the antinuclear movement and the major political organizations. Unlike in France, where the tradition of programmatic and ideological cleavages between the governing and the opposition parties allowed independence for dissenting groups, the German pressure for harmony and partnership tends to submerge them. Thus, as new groups sought to establish relationships with the SPD and the DGB, their attempts to maintain autonomy from political control became an important and persistent source of tension.

Just as in France the German student movement of the 1960s played a crucial role in establishing organizational, thematic, and tactical precedents for the ecology and antinuclear movements. The German Socialist students' organization SDS (Sozialistischer Deutscher Studentenbund) became the leading force of extraparliamentary opposition during the 1960s and through its criticism of the CDU-state transformed moral outrage into political protest. The student movement sought both university reform and more general political change, declaring solidarity with liberation movements in the Third World and opposition to fascist regimes. SDS contended that the authoritarian educational patterns that had once favored the rise of fascism still prevailed, that Germany with its consensus among oligarchical elites was a one-dimensional society, and that the Federal Republic was progressively integrated in the western military system as an ally of U.S. imperialism. The relatively wealthy and highly educated German youth of the 1960s rejected the conventional social climate of the CDU state and blamed the left opposition for having adapted to the evolution of a latent fascist society. Two political events associated with the economic recession of 1966 to 1967 encouraged such analysis: the significant electoral success of the neo-Nazi NPD (National Demokratische Partei Deutschlands) and the decline of significant parliamentary opposition following the big coalition of CDU/CSU and SPD. The students were able to mobilize mass protests against the constitutional and legislative changes that would allow police control and restriction of civil liberties in periods of national emergency.

To implement their long march through the institutions, many student activists joined the SPD, increasing this party's support in local and national elections but also encouraging internal party discussion of radical alternatives to capitalist economics. When the first social-liberal government took office in 1969, Willy Brandt was able to speak about a historical turning point. In fact the first years saw several significant changes: new policies toward socialist countries, educational reforms, environmental legislation, and efforts to democratize factories through co-determination. It even appeared for a brief period that extraparliamentary forces would be integrated into the traditional party system.

But the economic difficulties of the early 1970s, reinforced by the 1973 oil crisis, brought an end to the reform euphoria. The economic situation, combined with the climate of police repression during the years of terrorist activities, reduced political involvement and discouraged whatever hopes were invested in overall systemic change. During this period protests over local urban problems generated the citizen initiatives (see chapter 9), but unlike the student movement these new citizens' groups eager to attract a large constituency did not form an overall ideology. Nor did they seek a radical transformation of society.

Cultural conflicts in France and urban struggles in Germany both reflected the gradual evolution of public concerns toward questions of environment and lifestyle. But certain differences continued to prevail. French activists in a conflict-oriented culture are outspoken on political and ideological matters. Their ideological focus reflects the centralization of decision-making power: when nearly all local decisions are made in Paris, influence must be directed to the highest policy level. Unable to effect specific changes directly, opposition groups tend to articulate general ideological criticism. Lacking mechanisms for conciliation in areas of disagreement, they turn to dramatic action:

The virtual impossibility of bringing about any significant change on local or middle levels of the system engenders frustration and apathy about politics and support for political ideologies and parties that aim at the toal capture of the state and total transformation of the social system.[23]

In Germany, where people avoid identification with radicalism, political statements tend to be more cautious and pragmatic. The responses on surveys suggest that many even avoid expressing an opinion on nuclear power. Yet in the German decentralized political context the possibility of direct influence over decisions is far greater than in France. This has favored a more technical, empirically based approach that focuses more on specific problems than on overall societal transformation.

In both countries the new protest movements of the 1970s often tried to align with the left; but, disillusioned by the experience of the student movement, they gradually moved to establish autonomy from traditional political organizations. This evolution toward autonomy is most pronounced in Germany. In France the conflict orientation and the existence of opposition parties reluctant to adapt to the dominant political system have allowed some degree of protest to be expressed through existing organizations to a somewhat greater extent.

Finally, in both France and Germany nuclear power has raised issues that appeal to quite diverse groups affected by the dramatic postwar social and economic changes. It represents the industrialization of rural areas, the concentration of economic activities, and the centralization and depoliticization of decision making. It poses specific threats to agricultural life and environmental quality. It highlights the political impotence of traditional political institutions in the face of a closed, technical decision-making process. It exacerbates the frustration of young, educated groups disappointed with the outcome of the student movement and unable to achieve either the quality of life or the political control they envision. It has served as an ideal issue to bring together the diverse citizen associations that emerged in both countries during the early 1970s and were becoming increasingly frustrated in their attempts to bring about social change through legitimate political channels.

9 Organizations

The antinuclear movement provided an outlet for many people who had been politicized during the student movement of the late 1960s. It addressed the fears of small shopkeepers, artisans, and farmers threatened by technological progress or by land expropriation. It expressed the agendas of very different political groups. The initial evolution of antinuclear groups in France and Germany reflected the traditional role of voluntary associations in these countries and the willingness of established environmental groups to support controversial antinuclear activities. But as the organizations evolved, their structures converged. The movement became a network of local groups linked by their shared interest in stopping the nuclear program, their common literature and their active leadership.

The Tradition of Association

In February 1975 the prime minister of France appointed a working group of civil servants to reflect on ways to increase the participation of the French people in improving their quality of life. This commission chaired by Pierre Delmon diagnosed the situation and emerged in January 1976 with a controversial report.[1] It estimated that there were over 300,000 associations in France, multiplying at a rate of 22,000 each year. Interviewing leaders of these organizations, the commission found a high level of alienation and political frustration that suggested important inadequacies in the representative process. It proposed forty-five institutional changes to permit greater public involvement. The Delmon report was immediately classified, but associations of all types continue to proliferate, especially in the area of environmental protection.

By the late 1970s official environmental organizations (those registered and formally recognized by the ministry of the environment) included 250 national and 879 regional associations.[2] But in addition hundreds of local unregistered *comités* have formed around specific urban and environmental problems. Defenders of forests, guardians of suburban life, opponents of dams, highways, and airports organize to protest against nearly every government or industrial project. A major personality behind the organization of these small associations in the Paris region was Jean Claude Delarue, a professor of English. Disillusioned with the left, Delarue had resigned from the Socialist party at the end of the 1960s and began to mobilize middle-class suburban groups who—in the tradition of the petite bourgeoisie movements of shopkeepers in France—were concerned about preserving their rights. Seldom able to influence policy through direct political channels, the actions of these mini-groups were often militant: demonstrators squatted on a highway and created an enormous traffic jam to force the construction of an antinoise wall in Hayes-Roses, and they blocked the railroad tracks in Fontainebleau to get supplementary train service.

While defensive organizations are certainly not new in France, their proliferation is striking in the context of a long history of ambivalence toward voluntary association. Alsace with its strong German cultural influence is the only region where such groups have traditionally thrived. In the republican and jacobin tradition, the French have shunned associational life. The Napoleonic civil code prohibited the formation of groups of more than twenty persons without special authorization, assuming that they would restrict the rights of individuals and undermine the general interest. Following the revolution of 1848, the new republic decreed freedom of association, but less than one year later a law of November 27, 1849, withdrew this decree, and it was not until March 1884 that occupational associations could legally exist.

Official policy did not necessarily imply the absence of associational life; clubs, secret groups, and mutual aid societies formed despite legal obstacles. For example, in nineteenth-century France embryonic trade unions developed as mutual aid societies, Saint Simonists and Fourierists organized underground, and political organizations were disguised as religious groups.[3] Eventually in July 1901 France recognized the right of association. Under this law organizations may register with the local prefect if they place on file their membership list and a constitution that declares their intention to act in the public interest. Once registered, recognized interest groups may participate in formal consultation procedures.

The continued ambivalence toward association has been evident in the occasional attempts to amend the 1901 law. After May 1968, for example, the government proposed an amendment that would permit local authorities to delay recognition of a new association for up to two months in order to determine if it was formed for illegal purposes. The bill was judged unconstitutional.

The recent proliferation of environmental groups in France is all the more remarkable in light of the prevailing skepticism about citizen participation. Sociologists explain this ambivalence in terms of *l'horreur du face à face*, a tendency to avoid relationships that could lead to interpersonal conflict or a loss of individual freedom.[4] Perhaps more important is the limited confidence that group action can effectively influence a remote central government. The political reality of centralization and the weakness of local government discourage the formation of pressure groups.

The left has been especially dubious about voluntary organizations, regarding them as a manifestation of special economic interests. In the past the only significant encouragement of associational life came from the Catholic right, reflecting their political history since the Third Republic. The gulf between the political cultures of Catholicism and anticlericalism that opened during the French Revolution deepened after the establishment of the Third Republic and the 1905 law separating church and state. The intransigence of the republican regime isolated faithful Catholics into what has been described as a political ghetto. They responded by creating a flourishing network of private associations.[5] For Catholics these associations represented their corporatist hopes that a society based on small social units would replace the centralized republic. In its first years the Vichy regime seemed to materialize such hopes. But then, disappointed by the Vichy experience, some Catholic political activists joined the resistance movement.[6] Later in the postwar years some Catholic organizations, in particular the youth groups and the labor union, CFDT, turned toward the left where the ideology of *autogestion* appealed to their ideal of a society of human scale. The involvement of these groups had significant influence on the evolution of the French ecology associations.

In Germany the right of association was one of the most important demands of the 1848 Revolution. The constitutional assembly in Frankfurt devoted long debates to this issue, but the different German states varied widely in their restrictions over political and professional organization. The status of voluntary association changed several times following the 1871 unification of the German Reich under Bismarck. At first the states retained the right to regulate the right of association. In 1908 after a period of severe repression of socialist organizations common legislation recognized the right to form associations or societies for purposes not contrary to the criminal law. The Weimar Republic (1918–1933) inscribed this right into its constitution. Then the Third Reich (1933–1945) abolished all associations controlled by its political opponents and created its own national mass organizations as a means of controlling every aspect of social life. Some associations survived this period, in particular the large nature conservation organizations which had developed a nationalistic ideology in conformity with official doctrine. The right of association was again granted by the Allied forces in 1945 and written into the constitution of the Federal Republic in 1949.

During these times of rapid political change most associations were apolitical, organized around leisure activities. In the more stable postwar climate of the Federal Republic, political and professional organizations have proliferated, encouraged by the decentralized administrative system and the real possibility of influencing policy through the *Länder* and the courts.[7] Today the average West German is likely to belong to several formal organizations, often simply because membership can bring certain advantages. Traditional environmental associations, for example, offer inexpensive travel opportunities and this, more than ideological commitment, accounts for their very large size.

Private economic associations in Germany have also thrived, strengthened by their contribution to postwar recovery. About half of Germany's adult population belongs to some association representing economic interests. Noneconomic interests, however, have been less well represented, and these needs are what inspired the proliferation of citizen initiatives beginning in the late 1960s. The citizen initiatives formed initially as local community responses to public transportation and urban renewal policies. Between 1970 and 1973 they were most active in Frankfurt, with local citizens squatting to resist the demolition of residential districts to construct office blocks.[8] The political strategy of citizen initiatives, borrowed directly from the student movement, was to mobilize mass public support. Their growth reflected the latent dissatisfactions developing from problems of urban growth, declining services, and environmental deterioration.

As the citizen initiatives turned to environmental and social issues, they began to expand. By 1977 a survey estimated at least 50,000 initiatives with a total of more than 2 million members, significantly more than all the political parties together: an opinion survey found that only 12 percent of the population would consider membership in a political party, but 34 percent would join a citizen initiative, and 60 percent would join one if their interests were at stake.[9]

The antinuclear groups were part of this organizational syndrome. In both France and Germany nuclear power gradually became an important focus for citizen initiatives and many new political groups that organized exclusively to oppose this technology.

The Environmentalists

In nineteenth-century Germany environmental associations flourished as organizations of nature lovers, their popularity reflecting the high value placed on nature in the German cultural tradition. After World War II the associations merged into two umbrella organizations: the Deutscher Naturschutzring with 2.5 million members and the Deutscher Heimatbund with 500,000 members. Both groups exist to promote the appreciation and conservation of nature. They present themselves as apolitical, constrained by their status as quasi-official, tax-exempt, and subsidized organizations and by their constituency, which includes members of all political parties. Although they avoid political issues, pressure from antinuclear activists in their ranks has brought some cautious statements about nuclear policy. Without explicitly formulating a position, the official bulletin of the Deutscher Naturschutzring has published reserved comments about the limited attention given to the environmental impacts of nuclear policy. Similarly the newsletter of the Deutscher Heimatbund avoids a definitive position but criticizes specific plans and the secrecy of the decision-making process.

During the 1960s some associations began to address environmental policy questions. The Bavarian Bund Naturschutz, a conservative organization linked to the Bavarian political establishment, increased its political activities to inspire environmental legislation which is among the strictest in the Federal Republic. This Bavarian organization expanded into other *Länder*, and in 1975 five *Länder* groups coordinated to form the Bund Natur und Umweltschutz Deutschland, specifically to influence national environmental policy. They elected Herbert Gruhl, former environmental spokesman of the CDU in the *Bundestag* as the first president.

The *Bund*, with 35,000 members, is more politicized than most German environmental groups, integrating the theme of environmental protection with a global analysis of the catastrophic consequences of industrial policy. Its periodical *Nature and Environment* devotes considerable critical attention to the problems of nuclear power, and it published a special issue on the police brutality at the Brokdorf antinuclear demonstrations.

The growing interest in the environment also motivated some groups to revitalize a biology-based world view, promoting ideas close to those of Nazism. An example is the Deutsche Lebensschutzverbände, an active branch of the World's Federation of Life Protection (Weltbund zum Schutz des Lebens) created in Austria in 1960 with affiliates in thirty-four countries. The German branch has forty affiliated groups throughout the country, and most of its members are physicians or biologists. It publishes a periodical *Life and Environment (Leben und Umwelt)*.

Many small German organizations formed around the political concerns and frustrations of locally known personalities. In 1968 Hermann Spielmann and several lawyers organized the Association against Parliamentary and Bureaucratic Abuse (Verein gegen Parlamentarischen und Bürokratischen Missbrauch). Its newsletter *Heisse Eisen* publishes material on environmental issues and the political dangers of a plutonium economy. Then in 1977, when so many ecology and protest parties were forming, Spielmann launched his own "nonparty of mature citizens."

As environmental concerns gained priority on the policy agenda, industry promoted its own associations. Action Clear Landscape (Aktion Saubere Landschaft), for example, is run by a board of directors, including several ministers, the presidents of glass, sand, and cement industries, and the directors of several factories.

Most German environmental associations are financed by dues, but the larger tax-exempt general interest associations receive public subsidies for specific projects. Their members include prominent politicians, and their leadership is integrated into an environmental establishment in which they cooperate with ministry officials and industrialists seeking to influence environmental legislation. These ties constrain their involvement in the nuclear debate. While many environmental organizations have criticized aspects of nuclear policy, they avoid explicit positions that would jeopardize their influence in other environmental areas. Thus the antinuclear movement developed out of the citizen initiatives and gained relatively little support from the large, established environmental associations.[10]

In France the environmental organizations, less well established and with fewer ties to the political system than their German counterparts, were more receptive to political engagement. The French environmental associations originated in the scientific and conservationist concerns of the nineteenth century. The early nature protection movement, represented by the Société Impériale Zoologique d'Acclimatation,created in 1854 by Isador Geoffroy St. Hilaire, presented exotic species of the colonial world to the French public, promoted the conservation of natural resources, and encouraged tourism. Other groups devoted to wildlife protection formed in France in the early part of the century, but none acquired the mass membership characteristic of the German associations.

The Société Impériale, renamed the Société Nationale de Protection de la Nature in 1957, assumed a policy role, addressing such problems as river pollution and deforestation. In 1969 it became involved in a battle to save a national park in the Vanoise where real estate speculators planned to develop a ski resort. Government collaboration in the Vanoise development scandalized environmentalists. Several groups merged into a national federation for nature protection, Fédération Française des Sociétés de Protection de la Nature (FFSPN), and this became an umbrella organization with 80 affiliated associations and 100,000 members. The battle to save the park politicized a generation of environmentalists far more active and militant than the early conservationists. The event also turned the movement toward a socioeconomic and political analysis of environmental issues—an analytic approach that was to characterize the subsequent development of the French environmental movement.[11]

Compared to the German associations, FFSPN has a relatively small membership. It is independent of traditional political organizations and maintains few contacts with leading public figures. Its autonomy allows relatively unconstrained criticism but reduces possibilities for direct policy influence.

During the early 1960s two new national but small environmental protection groups emerged: Nature et Progrès, concerned with promoting organic agriculture, and Vie et Action, focusing on health and nutrition. Both link environmental issues to the dehumanizing features of industrial society and propose a more natural life as a cure for the evils of modernity.

Finally, as in Germany the environmental movement stimulated some very small groups of missionary individuals who tie environmental issues to broader themes of world government and peace. But in the French context these individualistic and often obscure groups have a quixotic style that differentiates them from their serious and often angry German colleagues. For example, to celebrate International Day of the Environment in June 1978, the staff of *Combat pour l'homme,* a periodical published by "a lonely fighter for the environment," organized a bicycle tour from Paris to Moscow. It seems that the very absurdity of trying to influence the policy process in France allows a certain humor absent in Germany, where failure to effect administrative change leads rather to despair.

Working more to arouse public consciousness than to influence the policy establishment, French ecologists could afford to be openly critical. In contrast, their German counterparts, seeking to integrate themselves into the political structure and exert direct policy influence, were necessarily discreet. In the same light the French could also afford more than their German colleagues to back the highly politicized antinuclear associations that gradually began to dominate the environmental scene.

The Political Ecologists and the Antinuclear Groups

After 1974 the nuclear issue became the major focus of the environmental movement, giving rise to the political ecologist and a proliferation of organizations with a radical image devoted primarily to nuclear opposition. Two characteristics distinguish this new type of association from the earlier policy-oriented environmental groups: an explicit rejection of nuclear power and a militant style. In both countries the new political ecologist was a grassroots activist, extending the goals and skills of the student movement to this new arena.

In Germany nuclear opposition began within the citizen initiative movement. Many of the initiatives were small, local, and temporary coalitions formed to deal with immediate problems; they disbanded when their mission was achieved. But the environmental initiatives, often engaged in long-term struggles, were more stable, and they often formed larger regional associations as their concerns required greater coordination.

In 1972 about 1,000 local groups created an umbrella organization, the Bundesverband Bürgerinitiativen Umweltschutz (BBU). BBU, run by a central assembly of representatives from affiliated organizations, serves as a communication and organizing center. It maintains contact with environmental experts, gives tactical and technical advice to its affiliates, and lobbies in Bonn. Its constituent groups remain autonomous, but BBU has become the political center for the antinuclear movement.

Initially most BBU's affiliates concentrated on specific problems of pollution, traffic, or industrial siting. Today they have multiple objectives, but their major concern is nuclear power. In fact affiliation now requires an environmental ideology that extends beyond opposition to a single project; one group formed to prevent construction of an airport was excluded from membership because of its pronuclear position.

The BBU includes people from a wide political spectrum, but its core support is from the left; its first three elected presidents came from the progressive wing of the Protestant church, from the Liberal party, and from the SPD. Although the leadership tries to maintain communication with traditional environmental associations, these more conservative groups shun BBU's activist style and collaboration with radical groups during antinuclear demonstrations. In the German political climate fear of identification with potential enemies of the constitution is a significant barrier to cooperation. Thus the major support for BBU comes from young activists, the intellectual left milieu, and local groups immediately affected by nuclear projects.

In France many of the political ecologists came directly from the environmental movement, but a new generation of local and regional associations formed in response to specific nuclear siting plans. In 1969 the first local antinuclear group, the Comité Bugey Cobaye, organized marches and circulated tracts seeking to raise public awareness of the nuclear issue. Similar *comités* formed at other nuclear sites, and by March 1972, when it was clear that this region would become a center for the development of nuclear power, they began to coordinate through the Fédération Rhône-Alpes de Protection de la Nature (FRAPNA). FRAPNA, with about 60,000 members, is an affiliate of FFSPN, illustrating the willingness of the traditional French environmental organizations to be associated with the nuclear debate.

One of the strongest regional antinuclear associations, the Comité de Sauvegarde de Fessenheim et de la Pleine du Rhin, formed in Alsace in 1970. In this region strongly influenced by German culture traditional environmental associations had long attracted a large membership. Here the church too plays an active role in environmental politics; indeed in 1977, 300 clergymen spoke out publicly against nuclear power. The Alsatian Association for the Protection of Nature, with some 90,000 members, helped to orchestrate the first ecology battles. But as the tactics of protest accelerated, unity within the organization could not be maintained. In 1976 the association split over tactical questions: the more conservative wing, convinced of the merits of lobbying elected officials, formed the Groupe d'Etude et de Concertation pour l'Environnement et la Nature en Alsace (GECENAL). The more radical members created the Association Fédérative Régionale de Protection de la Nature de l'Est (AFRPNE). The wisdom of mass demonstrations was the main contentious point. But as in Germany traditional political alignments also divided the environmentalists: many older members identified with the Alsatian political establishment, in particular the Christian Democratic CDS, whereas the younger, more radical environmentalists sympathized with the left.

The Alsatian ecology associations also attracted activists from the regional cultural autonomy movement who shared their concern about technological development and its implications for future local control. Just as ecologists defend the natural environment, so autonomists defend the social and cultural environment. Their active participation in ecology associations in Alsace, Rhone-Alps, and Brittany have helped to turn these regions into political centers for the antinuclear movement.

The formation of national coordinating associations is especially important in France, given her centralized political system. On the national level Les Amis de la Terre has been the most active group in the antinuclear movement. When formed in 1970 by people who had participated in the American Friends of the Earth, it had about 1,000 members, mostly Parisian intellectuals with a left-liberal background. Hoping to raise public awareness of environmental issues, they published a translation of the U.S. *Environmental Handbook* and a newsletter, *Le Courrier de la baleine.* They organized folkloric demonstrations such as a bicycle rally in Paris against the highway planned for the left bank of the Seine. The Parisian elitism of the early Amis de la Terre activists at first made it difficult to collaborate with the associations in the provinces. But the organization expanded very rapidly when it began to oppose the nuclear program in 1974. Many activists from the May 1968 student movement joined this association, which grew to include over 150 groups with about 10,000 members. It entered formal politics by backing René Dumont in the presidential campaign of 1974 and Brice Lalonde as an ecology candidate in a parliamentary election of 1976.

The association's internal organization reflects its ideology of decentralization and *autogestion.* Local and regional affiliates choose independent policies, although they must be compatible with those determined by the representative assembly of the national organization. Local groups raise their own financial resources from dues and book royalties and pay 10 percent of their income to support a national documentation center used as a common resource.

Les Amis de la Terre shares the theme of *autogestion* with the CFDT and the Socialist party, and several socialist intellectuals (the sociologist Alain Touraine, the ecological candidate René Dumont, and the journalist Michel Bosquet) have emphasized the need for closer ties. But most ecologists remain aloof from formal political associations and dissociate themselves from the traditional left:

This utopia, this desire for a less hierarchical, more open, more just society, this force growing out of socialism, is today invested in ecology. There is no future on the right . . . ; but neither do we consider ourselves part of a symmetric left. Certainly the historical failure of the left accounts for our existence. . . . It would be dangerous to consider the ecological movement as an heir of gauchism.[12]

Its neutrality exposes Les Amis de la Terre to criticism. Conservative critics accuse it of cooperating with the CIA and U.S. multinational corporations to reduce European competitiveness in the nuclear field. Critics from the left call it the CGT of ecology—a professional and technical union engaged only in reformist tactics and political compromise. The leadership of Les Amis de la Terre, however, maintains that political independence is necessary to attract wide public support and increase political influence.

Its independence has in fact attracted the support of some less politicized organizations. One of these is SOS-Environnement. After the 1976 municipal elections, several hundreds of the small middle-class initiatives that had been organized by Jean Claude Delarue created SOS-Environnement as an umbrella organization. Its agenda was to promote local democracy and greater access to administrative information. This group attracted a relatively apolitical constituency that carefully avoided the increasingly ideological nuclear debate. However, during the 1978 election campaign, SOS-Environnement accepted an antinuclear position for the sake of unity among environmental groups. After the elections it returned to its more ambiguous stance to preserve harmony within its own ranks.

As the antinuclear movement developed in France and Germany, the style of their organizations converged. Both Les Amis de la Terre and BBU are more social networks than formal associations. Permanently committed members are few in number, yet they have been able to mobilize a heterogeneous constituency to participate in massive demonstrations.

The Network

In October 1976 the German police expelled 500 demonstrators who were occupying the site of the proposed nuclear power plant at Brokdorf. In response, over 30,000 people descended on Brokdorf from all over Germany to express their solidarity with the antinuclear activists. They had been mobilized in less than two weeks. Every leading ecologist in Germany was there.

Despite ideological cleavages, apparent fragmentation, tactical failures, and the discouragement that followed violent clashes with the police, the antinuclear movement has maintained its ability to mobilize large demonstrations on short notice. But mobilizing its diverse constituency is a major challenge for the fragile antinuclear groups. The permanent core of committed activists is small; within each association no more than 10 to 20 people actually devote an important part of their time to organizational activities.[13] Yet this is no indication of their functional size; the strength of the organizations lies in their ability to convene a much larger constituency during key events—demonstrations, elections, rate-withholding, petitions, and other antinuclear actions. In essence the antinuclear movement is a form of intermittent social organization that oscillates between a small core of committed activists and a mass organization temporarily mobilized for special events.

The intermittent nature of the movement precludes a formal hierarchical structure. The movement can best be characterized as a network of diverse and autonomous groups sharing common values. A study of similar American grassroots movements describes their loose-knit, multicentered organizations as a "badly knotted fishnet with a multitude of nodes of varying sizes, each linked to all the others." [14] Similarly the local antinuclear groups, or nodes, that comprise the BBU or Les Amis de la Terre function with complete autonomy, but their leaders meet regularly and appear at all strategically important occasions. However, the major factor sustaining this network is the communication maintained by the active ecology press.

In France nationally distributed ecology periodicals thrived especially during the most active years of the antinuclear movement. The first ecology columns appeared in 1969 in the satirical publication *Hara Kiri,* a self-claimed "silly and nasty" newspaper, and *Charlie Hebdo,* a cynical, anarchistic, and often outrageous weekly with a circulation of more than 10,000. Highly critical of the value system of industrial society, *Charlie Hebdo* appeals primarily to a young generation of intellectuals. But its sensitivity to current social issues and expression of often repressed feelings about official politics also appeals to young technocrats. *La Gueule ouverte,* published by a group of radical ecologists and pacifists, provides both technical and political material on environmental, especially nuclear, issues.

Publishing houses also address environmental issues. The most active left-wing publisher in France, Maspéro, printed Laura Conti's analysis linking ecological issues to the class struggle *(Qu'est-ce que l'écologie? Capital, travail et environnement)* and J. P. Colson's analysis of the undemocratic nature of French decision making in nuclear policy *(Le Nucléaire sans les français)*. Jean Jacques Pauvert publishes a series by Les Amis de la Terre. Among the bigger houses Le Seuil and Stock devoted several books to the nuclear issue. Ecology groups also circulate photo brochures and case studies to document their struggles.

In 1973 the left liberal weekly, *Le Nouvel observateur,* asked one of the founders of Les Amis de la Terre, A. Hervé, to prepare a special ecology issue.[15] Following this special issue, *Le Nouvel observateur* helped launch *Le Sauvage,* with a circulation of some 25,000 and an editorial staff largely composed of members of Les Amis de la Terre. Finally, as environmental themes assumed growing importance in the mass media, a specialized press agency formed in 1973 to publish a monthly called *Ecologie* and a weekly called *Apri Hebdo.*

Ecologists in Germany have not developed a national ecology press. Instead the citizen initiatives in different regions are linked through newsletters that report their activities. In addition socialist intellectual groups have helped to mobilize support for the antinuclear movement, by giving priority to ecological themes in their publications. The Sozialistische Büro in Offenbach edits a monthly bulletin, *Links,* that discusses the political aspects of the environmental movement. The independent left periodical *Kursbuch,* edited by Hans Magnus Enzenberger, devoted several issues to ecology and citizen initiatives. Freimut Duve, member of the SPD, founded a periodical *Technologie und Politik,* which publishes critical articles on environmental and nuclear politics. Several periodicals, *Konkret, Pardon,* and *Dasda,* directed toward JUSO and other groups in the SPD provide information on the activities of the political ecologists.

Several publishers have focused on the nuclear debate. The Hamburg publishing house, Association, run by regional citizen initiatives, prints their technical dossiers and documents their actions. The Verlag für das Studium der Arbeiterbewegung, a publisher for the labor movement, developed a book series on the citizen initiatives. The Berlin house, Wagenbach, printed the text of a movie on Wyhl, *Better Active Today than Radioactive Tomorrow!* Among the larger German houses, Rororo devoted several issues of its series *Aktuell* to the nuclear debate. Fischer created a new collection *Alternativ* devoted to ecological and antinuclear issues that relate to new lifestyles. Luchterhand and Pahl-Rugenstein focused on the political aspects of the nuclear debate. Finally, the major liberal magazines *Der Spiegel* and *Die Zeit* report on nuclear energy problems in nearly every issue. This openness of the intellectual community to the antinuclear issue has amplified the influence of the political ecologists and created ties between their dispersed organizations.

Several international organizations also help to maintain coordination among the loose network of groups that comprise the antinuclear movement. The European Environmental Bureau established in Brussels in 1974 serves as an umbrella organization and an information exchange for forty environmental associations from nine countries. In 1977 it sponsored an international conference on the problems of a nuclear world.

The Worldwide Information Service for Energy (WISE) was founded by the Smiling Sun Foundation, a group based in Amsterdam and sustained by the royalties from the symbol used on posters, T-shirts, and bumper stickers. WISE publishes an international bulletin with press releases about antinuclear actions, runs an information service, and facilitates contacts across national borders.

The leadership of the antinuclear movement is perhaps its most important source of coordination. But ironically dependence on leadership creates the very pattern of hierarchy and centralization that the movement ideologically abhors. Media attention focuses on the leadership, adding to its power and visibility. Laws require legally recognized voluntary associations to have officers able to speak for their constituents, encouraging hierarchy and formal structure. The growing internationalization of the movement gives prominence to leading representatives, further concentrating their power. An active leadership is a source of strength in a diffuse and intermittent social organization, but it is also a source of strain.

The persistent ideological factions within the movement contribute to strain. French associations face frequent sectarian disputes within the left; their intellectual leaders constantly debate the correct definition of such concepts as *autogestion* or revolution. In Germany social strains appear within local initiatives. In Wyhl and Brokdorf farmers with conservative politics found themselves fighting nuclear power alongside young activists with radical political goals. Local occupational groups with specific, pragmatic concerns found themselves aligned with veterans of the student movement who had become professional activists. Different ideologies demand different slogans, symbols, and strategies. Which groups should march at the beginning or at the end? Should each group emphasize its own symbols? Should there be a common leaflet or separate ones? Should red and black flags be allowed?

Such questions are especially serious in Germany where a display of radical symbols could harm the general cause. But in France as well, the anti-nuclear movement must maneuver among diverse interests. For conservatives it must emphasize the legitimacy of its organization, arguments, and strategy with respect to the constitutional political order. For more radical activists it must emphasize the potential political transformations that might eventually occur. The fluid and diffuse network organization of the movement is fragile, but it allows activists to sustain the interest of a broad constituency and address the very diverse ideological concerns that move people to oppose nuclear power.

"Nuclear Energy Kills" (Courtesy of
Secrétariat National du Mouvement
Ecologique)

"Struggle for Life"

"Nuclear Energy Creates Jobs" (From *Europa Forum*)

"Atomic Energy: Necessary Evil or Planned Waste?" (From *Europa Forum*)

"Nuclear Society, 'Society of Cops'"
(Courtesy of Secrétariat National du
Mouvement Ecologique)

"Summit Meeting to Assess Objectively the [Three Mile Island Accident] Situation." Jimmy Carter, Giscard d'Estaing, Prime Minister Raymond Barre, and Chancellor Helmut Schmidt: "One Cannot Stop Progress" (From *Le Canard Enchaîné*)

10 The Discourse

How is one to interpret the rise of the antinuclear movement—its broad appeal and persistence even in the face of frustration, government repression, co-optation, and in France consistent failure to influence nuclear policy? A key to its appeal lies in the ideology expressed by its most active and articulate groups. Our analysis seeks neither to establish the validity nor to point up the fallacies of this ideology. Rather it deals with the antinuclear discourse as a strategic tool by which activists try to recruit and mobilize followers around a vision of reality and a set of broad social goals.[1]

At first the antinuclear discourse concentrated on the risks of the technology and its relation to a future ecological crisis. As the movement mobilized a larger social constituency, the activists deepened their analysis, gradually evolving a world view—a perspective on social change, a diagnosis of technology and its political consequences. The elaboration and propagation of a broad ideology transformed the antinuclear movement. Far more than simply a response to an immediate technological choice, it expressed a deep fear of disaster and a profound sense of crisis concerning modern industrial society and the character of its political life. It reflects the eclectic character of contemporary popular culture:

Today the images and the language of the ecologists . . . come from everywhere: from religion and Marxism, from the most plain scientism and the most incandescent mysticism. . . . You will find traces of Rousseau . . . Marx, Kafka, Charlot, and the Beatles. . . . We find our models of a future world in movies: *General Line* by Eisenstein, *Modern Times* by Chaplin, *A Nous la liberté* by René Clair, and *Yellow Submarine* by the Beatles.[2]

French activists refer to their *nébuleuse écologique,* their vague and confused environmental ideology. Devoted more to mobilizing people than to establishing an all-embracing organization, the ecology activists are careful to avoid too rigid a theoretical agenda, for while an ideology can unify a socially heterogeneous constituency, it can also establish criteria that would exclude certain groups. The pamphlets disseminated by the movement give the impression of product differentiation; they expound diverse themes depending on the public addressed: agricultural problems are stressed for the farmers, occupational hazards for workers, the dominance of large nuclear firms for middle and small businesses, the role of capitalism as the driving force behind nuclear expansion for leftists, and the destruction of traditional social structures for conservatives.

Such ambiguity is in some ways counterproductive; it makes established political groups such as socialist organizations and unions uneasy, preventing a dialogue that ecologists often wish to cultivate. But while ambiguous, the discourse conveys powerful themes: an apocalyptic image of the destructive potential of nuclear power, a pessimistic vision of ecological, economic, and cultural crisis, and a critical analysis of its sociopolitical roots. Opposition to nuclear power becomes a struggle for human survival and political justice. "Centralization," "nucleo-technocracy," "secrecy," "electro-fascism": such words pervade the antinuclear discourse. In this broader context it is sufficiently powerful to prevent major splits within the movement despite the very different political agendas and utopian images of its constituent groups.

Apocalyptic Imagery

Nuclear power is above all a symbol associated with death and war, and this is what drives the nuclear debate. Indeed the power of this symbol is the main force maintaining the unity of the antinuclear movement. Fears associated with the first use of nuclear energy—with Hiroshima and Nagasaki—persist despite all efforts to not dramatize the debate. Nuclear advocates use linguistic tactics to dissociate civilian from military uses of atomic energy. In the late 1950s "atomic" power became "nuclear" power. Nuclear industrial areas became "parks," and accidents became "incidents." In Germany official documents used the concept *"Störfall"* (irregularity) rather than *"Unfall"* (accident).[3] But new labels failed to remove deeply rooted images.

The association of nuclear energy with war, however, is by no means a direct one. Ironically antinuclear brochures of the 1970s seldom deal with the theme of nuclear weapons. Except for *La Gueule ouverte* in France produced by ecologists from the pacifist movement and several German publications from Maoist groups concerned about preparations for a nuclear force, the antinuclear press has generally neglected military questions.[4] On two occasions, however, military issues did gain prominence. In 1977 when President Carter announced his nonproliferation policy and pressured European governments to cancel nuclear exports, ecologists responded with a favorable press, although they later abandoned the issue of nonproliferation when European officials criticized Carter's policy as a means to reduce European competition on the world market. The antinuclear press also vigorously opposed the neutron bomb and the notion of a weapon that could kill people while leaving property intact.

Although seldom an explicit theme the image of war has implicitly dominated the nuclear debate, shaping the discourse even of nuclear proponents. The information circulated by the nuclear industry specifically articulates the differences between a nuclear plant and an atomic bomb. Seminars organized by EDF for its personnel have focused on extreme cases of risk—a plane crash or military attack on a power plant.[5]

Antinuclear activists exploit macabre associations in their arguments and imagery. The ecology press describes nuclear plants as quasi-weapons, threatening an invisible, slowly operating death.[6] A poster portrays a map of Paris within a skull ripped apart by the three rotary blades that form the symbol of radioactivity: "Nuclear power is at your door." Another poster shows a workman hammering nails into an endless row of coffins: "Nuclear energy creates jobs."

The cover of a popular book of photographs of the Brokdorf demonstrations (selling 20,000 copies) shows a mushroom cloud growing out of a cooling tower. The motto of the French ecology group Ecology and Survival is "Struggle for life." The slogan of a German citizen initiatives in Wyhl, and the title of their film, is "Better active today than radioactive tomorrow." Caricatures abound: a man in uniform equipped with gas mask and Geiger counter stands before a power plant surrounded by dead trees and mountains of garbage. A cow with a glowing udder has a malicious smile. Nuclear forces appear as skeletons, while the antinuclear symbol is of a smiling sun: "Nuclear, no thanks!"

Fears are reinforced by secrecy, for what is secret is suspect. Industrial secrecy, lack of information, and classification of documents on nuclear safety feed suspicions about claims that nuclear energy is safe and clean. The very nature of the technology perpetuates an aura of secrecy. The risk of contamination, after all, is unpalpable and dangerous; radiation is invisible and mysterious, and its consequences operate slowly and irreversibly. A "Poem of the 1980s" describes these invisible but pervasive risks:

Wait, wait a little while,
Strontium will come to see you,
In very small, refined particles,
In vegetables, in porridge and in beans.[7]

The apocalyptic vision of nuclear power extends to the image of a future nuclear society as a police state. After the May 1968 revolt in France and the terrorist wave in Germany police were equipped with modern materials and weapons, small tanks, water cannons, tear gas, and chemical defensive weapons. Antinuclear demonstrators wore heavy shoes, helmets, and masks to resist police truncheons and tear gas. The impression was less of protest than of war. The ecology press described the use of police in the 1977 demonstrations as a first step toward a repressive nuclear society, and the mass media spoke of civil war.

Apocalyptic visions preclude apathy, and powerful imagery forces a response. The ecologists directed this response by linking the nuclear issue to broader economic, political, and cultural problems—problems that appealed to a variety of social groups, but the persistent issue of survival remained a source of unity, preventing potential conflicts among an increasingly heterogeneous constituency.

Crisis

To ecologists, industrial society has arrived at a critical turning point:

A third of the animal and plant species existing at the beginning of the century have disappeared; a second third is close to disappearing; inevitably humanity itself will be part of the last third.[8]

For radical and conservative ecologists alike, nuclear power has become a symbol of crisis. Contrasting nature with civilization, the natural with the technical, they address widely shared concerns not only about the environment but about natural lifestyles and the crisis in human relationships. The interpretations of crisis, however, are mediated by different world views, political ideologies, and conservative as well as radical cultural traditions. In both cases ideological elements that had been forgotten during the postwar period have re-emerged, raising once again questions about the meaning of industrialization and modernization and the role of the state.

German ecologist Herbert Gruhl questions the very concept of progress symbolized by nuclear power. An acute sense of imminent crisis pervades the discourse of his following among conservative environmental groups. Their view reflects the pessimism that often invades the cultural mood in times of recession or after periods of rapid social change. Its moral tone recalls the great debates of the late nineteenth century when humanists reacted against industrialization and its effects on social life. Just as Ferdinand Toennies distinguished the depersonalized *Gesellschaft* (society) from the more organic and personal *Gemeinschaft* (community), so ecologists idealize the contrast between urban and rural life. An Alsatian group fears "we will soon no longer recognize the soil on which we were born." Farmers in Wyhl fought for their land: "Our homeland is our mother. Those who take our land away are killing our mother." Just as Max Weber saw rationalization as the disenchantment of the world, so ecologists of the 1970s argue that progress implies a mass society that eliminates diversity, centralization that breeds alienation, anonymity that destroys personal relations, and politics that reinforces the status quo.

At the beginning of the century the strongest expression of such views came from the German Jugendbewegung and the Wandervogel. Emerging after a period of rapid industrialization and urbanization, this youth movement attacked the evils of civilization and praised the values of a natural life. It deplored the atomization created by developing technology —the dissolution of organic relationships, the loneliness and ugliness of large cities, and the lack of aesthetic values in the industrialized landscape. It idealized community, nature, and a healthy country life.[9] Today the literature of conservative ecological groups in Germany (such as Bund, Weltbund zum Schutz des Lebens, GAZ, AUD) reads like a document from this early period of industrialization. Progress contradicts the laws of nature and industrialization and urbanization are the source of ubiquitous social problems: destruction of the family, psychological stress, prostitution, alcoholism, and drug abuse reflect the demise of rural life:

People everywhere try to get out of dirty agricultural work with its calloused hands and into white-collar professions in heated and air conditioned offices; to get out of rural insecure life with its good and bad harvests and meager years, and into the welfare state that protects against the hazards of life. . . . At first they destroy the village school and its autonomy, the local craftsmen followed, the meeting places and pubs were next; sometimes even the church together with the priest. . . . To be a peasant and a peasant woman must again become desirable. . . . The health of the soil and animals must be a primary objective as it was the way of peasants for centuries.[10]

Reviving the traditional concern about man's place in nature, environmental groups warn that all will be finished unless man abandons pretensions of biological superiority to become a partner of nature.

In France the ecological press conveys a less nostalgic tone, and its diagnosis of the crisis is far more political in style. This too has historical precedent. Conservative ecology groups often inspired by Catholic social philosophy base their critique of contemporary social trends on traditional corporatist ideals. Early in the century Catholic political groups, opposed to the anonymous organization of the secular republican state, conceived of a society organized around small, self-governing social units. Emmanuel Mounier, founder of the periodical *Esprit,* expressed this ideology as personalism, hoping it would replace the materialism expressed in both capitalist liberalism and the Marxist concept of class struggle.[11]

Personalist themes persist, especially in the ecology movement; they appear in the writings of Christian Democratic ecologists like Philippe St. Marc who believe that ecology is a revival of humanism in response to the moral decadence of capitalist and socialist societies. St. Marc portrays twenty-five years of urbanization as a social disaster:

The city, formerly a hospitable community, is a conglomeration of evil: places are crowded, housing is defective and located far from work, traffic is exhausting, green spaces are under cement. . . . Individuals are isolated in an anonymous mass. . . . In this psychological desert there rests only one escape: violence, pornography, or drugs. Submitted to the logic that everything must be profitable, they are themselves turned into sources of profit.[12]

Crime, juvenile delinquency, prostitution, divorce, and the demise of the family are thus perceived as the inevitable result of urbanization. St. Marc calls for protection of the family and recognition of the profession of mother as part of his ecological program. From this moral perspective nuclear power appears in the conservative ecology literature as a further social disaster likely to contribute to anonymity, disruption, and despair.

The cultural pessimism and ascetic moralism of conservative ecologists worries the left; such views, it is feared, can produce deep resignation or at best only mild reform.[13] But while avoiding moral stereotypes, the more progressive groups use quite similar images to convey a pessimistic view of a civilization in crisis. An editorial in *La Gueule ouverte* describes Paris as

A hypertrophic city, like a sick head, a city that will die in the same way as New York, a city that will explode in the same way as the social security budget, a city that will be livable only once the state has collapsed (read your Marx).[14]

Like their conservative colleagues many left ecologists reconstruct an image of a utopian past and deplore the dehumanization and decadence of modern life.[15] Indeed in this respect they often appear to converge with reactionary groups. However, these antinuclear activists, identifying themselves as politically left, blame the crisis less on individual materialistic attitudes than on the structure of capitalism. They talk less of moral decadence than of Herbert Marcuse's one-dimensional society, less of urban and environmental problems than of their political implications.

Yet these ecologists depart from the traditional left in significant ways. In particular they focus on conflicts outside the sphere of production, seeking alliances less with labor than with groups engaged in "secondary struggles." For example, opposition to political centralization has forged a tactical alliance between antinuclear groups and regional and cultural autonomists. Slogans are scribbled on the walls of southern France: "We are fed up with being governed by Paris." The Occitanie press actively supports antinuclear activities in the Rhone-Alps. Autonomists in Brittany include nuclear issues in their struggle against French economic and cultural hegemony:

To extend its economic influence, to find a docile labor force . . . , the French bourgeoisie fought against the popular culture. . . . They had to convert the workers of Brittany, Corsica, Occitania, and the Basque country into "good average Frenchmen" . . . , who, intoxicated by the bourgeois ideology diffused through mass media, would consume all the products of civilization, from cars to color TV sets and men's "low-cut" fashion underwear. Today the cultural resistance is expanding to become more and more radical along with the increasing struggles . . . against economic exploitation, . . . and the destruction of the landscape by a luxury tourist industry, military bases, and nuclear power plants. . . .[16]

Similarly the identification of crisis with political authority and social control has linked the antinuclear movement to feminists, who see the ecological crisis as part of a man-made society.[17]

In Germany citizen initiatives in the regional elections linked antinuclear and ecology issues to social exploitation; homosexuals, prisoners, the elderly, and unmarried mothers joined in a common ecology platform. The colored list in Hamburg and the Alternative in Berlin proposed a political project: to build an alliance of marginal groups against the anonymous agencies of the state. Their shared values have in fact led to an embryonic political front of marginals, as envisioned by philosophers of the new left such as Herbert Marcuse, Jürgens Habermas, and Henri Lefebvre.

Finally, the concept of crisis has inevitably associated the antinuclear movement with a youth culture that has expressed its political alienation in different ways. While the generation of activists socialized in the political climate of the late 1960s believe in the need for a minimal degree of organization and recognize some merit to traditional political behavior, an even younger generation is often disillusioned with all political action. Some seek a countercultural retreat from social activism and concentrate on individual emancipation or metaphysical preoccupation.[18] Others seek direct action or counterviolence and on several occasions have disrupted demonstrations. Both tendencies were visible in 1978 and 1979 when thousands of young people gathered at a meeting of alternative groups in Bologna, Italy, at a *Tunix* (do nothing) festival in Berlin, and at a Community Network meeting in Vienna, Austria. There urban Indians, delinquents, exprisoners, and ecologists demonstrated their solidarity and hostility to prevailing social values.[19] These youths, often called autonomists, are not a significant part of the antinuclear movement, but the similarity of their discourse defined by common perceptions of crisis creates inevitable associations that have been problematic for the movement's political image.

In both France and Germany the perception of crisis forged useful alliances between ecologists and diverse marginal groups, but these alliances also compounded the difficulty of communicating with the organized left. In France they brought recollections of the anarcho-syndicalist tradition of the 1920s and 1930s—a tradition actively opposed by the communists. In Germany the new marginal groups antagonized the left by debunking the integration of working class organizations into the capitalist system. Such tensions forced the movement to develop its own political diagnosis in far more explicit terms. As this diagnosis evolved, the ideological differences between conservative and radical activists began to assume increasing importance.

Political Diagnosis of the Presentation of the Enemy and Self

Just as ecologists juxtaposed urbanization against nature, society against community, so they presented themselves as a rising and constructive political force opposing the anonymity and hierarchy of established political groups. Moderate as well as radical ecologists share a common socioeconomic analysis—that the enemy, the source of environmental problems, lies in the growth mentality and profit motive held by government and industry alike. The ecology press, in its critique of nuclear power, emphasizes the complicity among government agencies and nuclear suppliers, utilities, and banks.

While government and industry see nuclear power as a rational means to enhance national goals, its critics see electro-fascism, the emergence of a social order in which institutions of repression and social control (the army, the police) are mobilized to guarantee energy production and economic growth: "The nuclear society is a society of cops." [20] While the establishment sees centralization and concentration as essential to imple-

ment energy policy, critics see technocracy developing at the cost of democratic values. They predict "the disappearance of all agencies of mediation
between the individual and the state. The nation is an enormous uniform
and shapeless body . . . that dominates through a bureaucracy with
thousands of strangling hands." [21] They portray the citizen as but a consumer of politics elaborated and executed at the center.

German ecologists blame the demise of democratic institutions on the
politics of power and profit. Sometimes the frustrations of democracy are
attributed to selfishness:

Today's institutions and ideologies are so selfish that, faced with a choice
of sacrificing themselves or humanity, they will decide the latter; and
this despite the fact that their own survival depends on the survival of
humanity. [22]

Sometimes they are attributed simply to bureaucratic behavior:

Administrative behavior has changed with automatization. Computers
rather than civil servants communicate with citizens. As the bureaucracy
becomes inhuman no wonder people react. [23]

Critical of political institutions, ecologists present themselves as a moral
force, a *courant d'idées,* and in some cases as conveyers of pure truth. [24] They
claim to represent those who cannot speak out in their own interests: potential victims of nuclear accidents, future generations, animals and
plants, and the Third World. They see themselves involved in a dramatic struggle for survival: "[We] are here to defend the last remainders
of our country's soil where our ancestors have lived for generations." [25]
Paradoxically the most conservative ecologists convey this image of a holy
war.

The more radical activists present themselves in a less dramatic light.
Heir to the antiauthority and anti-institutional ideologies of May 1968,
these ecologists distinguish themselves from the orthodox left. In particular they criticize the Communist party and the traditional labor organizations as oligarchic, neither controlled by their constituents nor representing their interests. [26] After endless discussion about their relationship with
the left, French ecologists in the most active regional associations and in
Les Amis de la Terre ultimately have opted for political autonomy.

In seeking to redefine their strategy, some ecologists refer back to earlier
traditions; others emphasize their new and original approach. Political
thinking among conservative ecologists recalls the attempts by French per-
sonalists and German conservative revolutionaries [27] of the 1920s and
1930s to formulate a "third way" beyond both capitalism and Marxism:

Capitalism, Marxism—same struggle. Beyond their divergence on redistributing the results of growth, how they converge on growth itself! They give the same priority to material goods over immaterial values, to the level of living over the quality of life. They represent the same domination of man over nature; the same religious glorification of science and technology; the same type of development based on concentration, gigantism, industrialization, and urbanization; the same supremacy of technostructure over the grassroots individual.[28]

While conservative ecologists seek a third way through reactionary models, the much larger left dominated by a younger constituency seeks to go beyond the right-left cleavage by defining a new political identity. Brice Lalonde locates the movement somewhere between radical socialism and left liberalism, identifying it as a third force in European politics.[29] But such efforts to define a new political identity are divisive. In France, where a left strategy, excluding the Communist party, has for several decades been inconceivable, the ecologists' program critical of state intervention created a political dilemma. The ambivalence among ecologists concerning electoral participation and their oscillation from politicization to denial of politics are divisive even among those radical ecologists who appear to be the closest allies.

Utopian Images

The ideological vision of the ecologists seeks to transcend left/right political alignments, but in some ways it represents a convergence of these traditional political views. Social Democrat Erhard Eppler and former CDU parliamentarian Herbert Gruhl seem to have more in common with each other than with their respective party leaders. The traditional Philippe St. Marc appeals to exstudent activists from the radical left. They all share utopian images of a future society based on strong communities and small social units. But conservative and radical ecologists have very different images in mind when they employ these concepts.

Conservatives in the antinuclear movement, with their traditional frame of mind, look back to the past and seek reforms directed to smaller production units, craftsmanship, agricultural revival, and rehabilitation of the family. Some conservative groups present an ambivalent economic program colored by both socialist and middle-class corporatist elements. The German AUD asks for the nationalization of corporations and banks, the development of a cooperative sector, and the encouragement of agriculture and small business. The party GAZ seeks an equilibrium economy with stable prices and wages, the harmonization of the economy with the laws of nature, the export of adapted technologies to less developed countries, and more recognition and justice to mothers, "the most important stratum of the population." [30]

The French CDS presents a similar program. The utopia of Philippe St. Marc is based on revival of spiritual Christian values in a society of affective communities, where people can differentiate between good and bad and accept obligations toward themselves and others. "To revive the sense of duty is to teach respect of others." [31]

In contrast, the program of the left ecologists developed as a challenge to the organized left. They questioned whether a socialist transformation through the conquest of the state (embodied in the strategies of both social democrats and communists) would resolve the problems they see as most salient. Moreover, disillusioned with centrally implemented reforms, radical ecologists broke with the orthodox left to elaborate a program of decentralized decision making, small social units, grassroots power, and direct democracy as a means to achieve redistribution of resources and develop countervailing sources of power. To these ecologists a nuclear society would obstruct this program. This is indeed the basis of the antinuclear position articulated by SPD's youth organization JUSO and the French CFDT.

For the more radical ecologists Illich's concept of conviviality provides a popular model of liberation and emancipation. [32] His utopian thinking is influential within the subculture disaffected with the limited change that followed May 1968. These groups seek alternatives to the poverty of consumerism, a poverty they define in terms of social and emotional relations. Their utopia has found expression in popular culture. A French movie *L'An 01* describes the wish to start again in a convivial society free from power relationships. A popular book, *To Work Two Hours Each Day*, documents the burdens of work. [33]

Biological images also pervade much of the new eco-science fiction literature describing a society that respects the fundamental biological equilibrium. The self-regulation of natural processes becomes a model for social organization in Ernest Callenbach's *Ecotopia*, very widely read in Europe. Similarly the well-known French biologist Henri Laborit argues that *autogestion* is the social organization that is best adapted to nature. [34] These images, with their environmental and social implications, have wide appeal. But the pervasiveness of biological models of harmony and equilibrium also provokes skepticism among the more politicized groups, who associate such notions with fascism and prefer to emphasize the class character of society that lies at the source of environmental problems.

The shared vision of a decentralized society with smaller social units and a more natural lifestyle helps to unify the ecology movement. However, some members of conservative groups such as Bund and GAZ argue that the environmental crisis calls for immediate measures of vigorous and centralized state control. Professor Dr. Heinz Kampinski, director of the Bochum Observatory and Institute for Space Research, describes an ecological society as a spaceship in which survival depends on the most stringent discipline:

In a spaceship astronauts have to adapt their behavior to the very strict rules of a system. They use their resources cautiously. They fulfill their respective functions with almost 100 percent precision. Everybody treats his neighbor as he would treat himself. . . . On Earth everybody lives without noticing his neighbor and certainly without respecting him; . . . such behavior is unacceptable in a spaceship; it would not function. Our behavior on the planet Earth can be compared to a spaceship in the larger cosmos and has to adapt rationally to the existing resources. Nobody can be allowed to disturb the system as a whole.[35]

Some ecologists claim to hold the truth and feel that in times of crisis this truth must be imposed even at the cost of democratic values:

Minorities who tell what is necessary about systemic ecological relations are right. They must prevail when parliamentary majorities and the electorate cannot decide.[36]

A farmer describes the environmental crisis as a national emergency that requires powerful and dictatorial government control:

We must all restrict ourselves. The rich, the politicians, the industrial leaders must provide the example. They must live according to Prussian frugality, to the modern Chinese model, or to Christian ascetic behavior. With revolutions and strikes our situation will become even worse. We need a forceful government of national emergency. Only such a government is capable of improving our difficult solution.[37]

Ironically the book *Communism without Growth* by the East German philosopher Wolfgang Harich appealed to conservatives.[38] Harich argues that the environmental problems can only be solved by dictatorship and that therefore socialist countries are better prepared. His book, published in the Federal Republic in 1975, was never released in East Germany, and he left his country in 1979 for Austria.

These ideas, recalling classical German right-wing thought, form but a small part of the antinuclear discourse.[39] The mainstream literature rejects the notion of an eco-technocracy. Indeed most ecologists fear a green *goulag:*

Listening to our eco-technocrats and eco-philosophers, I imagine that in less than twenty years we will be led by a new elite possessing a global view of all interactions in the universe and proposing a society of material and mental rationing. As compared to this future, our present society looks like sweet paradise.[40]

In a fictional interview the leader of an ecology party of 1984 worries about the authoritarian implications of environmental politics:

Question: What do you think about the Party for Survival created by former center party deputies?

Answer: The Party for Survival mobilizes public fear in face of increasing pollution and nuclear proliferation in order to come to power. This is a very skillful but dangerous maneuver! It assumes that individuals are incapable of organizing and solving their own problems. Under such conditions, a coup d'etat . . . would be of no surprise! An accident at a nuclear power plant or any other pretext could justify the worst dictatorship![41]

In its images of the future the diversity within the antinuclear movement is clearly revealed. For its more conservative participants nuclear power and environmental degradation are moral issues. Concepts of sin, values, ancestral rights, our country and our soil, discipline, restrictions, austerity, and savings are sprinkled through their discourse. For others— the young political ecologists toward the left of the spectrum—political and social images, *autogestion,* eco-utopia, and the quality of life, prevail.

For still others, less established and with a more cynical view, nuclear power is but another modern irony. Avoiding global solutions, they seek a life of immediate reward: pleasure, desire, imagination, emancipation, and leisure color their dialogue.

For each of these groups a vision of a different future provides a sense of historical mission. More than simply opposing a technology, ecologists foster through their discourse the conviction that they can change the rules of the political game, that they can represent the forces of tomorrow and transcend the political alignments of today.

IV THE CONTAINMENT OF CONFLICT

11 The Role of the Courts

The 1976 judicial decision to block the construction of the nuclear power plant in Wyhl was a key event in the European nuclear debate. Ecologists throughout Europe applauded this decision as a demonstration of how citizens with few resources could resist industry and the powerful bureaucracy of the state. Indeed ecologists in both France and Germany regard the courts as one of the more important means to influence nuclear policy, and since 1973 have brought nearly every siting decision to court.

Yet the results of legal action have been completely different in these two countries. While German environmentalists have successfully used the courts to bring the government's nuclear energy program to a standstill, the legal appeals of French ecologists have had virtually no effect. For French activists the courts are a means to create publicity, impose delays, and increase the cost of nuclear power. "There is no reason to be very optimistic about the scope and efficacy of the legal struggle against the present nuclear program—at least in the present state of nuclear regulations and their interpretation." [1] In contrast, German activists see the courts as a direct means of public control over nuclear power. "Future discussions about nuclear energy will depend on judicial decisions. Until now the courts have approached the issue of nuclear technology only in a superficial way; but later the problem of the industrial use of nuclear energy will have to be judged according to constitutional principles." [2]

The responsiveness of the courts arises less from the tactics used by ecologists than from certain characteristics of the legal structure in each country, and especially from the relationship between the judicial and political system. As a decisive factor in the evolution of the nuclear power controversy, the courts, their political relationships, and their decisions merit special attention.

Legal Structures and Legislative Mandates

The legal system in France and Germany is based on Roman law. This system differs from the common law traditions of the United States and England in several ways that are important for understanding the role of the courts in the nuclear debate. [3] First, in dealing with cases involving the administration, the courts usually define themselves more as defenders of the state and the general interest than of individual rights. Second, class action suits are not admitted in Germany, and in France they may be heard only if specifically authorized by law. This gives environmental associations relatively few opportunities to sue. Third, Roman law justice is intended to enforce, not to interpret the law. Interpretation is normally an administrative prerogative. With respect to nuclear power, however, the atom law in Germany opened possibilities for judicial interpretation. But in France with no special legislation, the courts remained confined to litigation over legal technicalities and procedural breaches of the law. Finally, precedent is not necessarily binding on court decisions in the Roman law tradition. Faced with the same substance, different courts have come out with unpredictable and even contradictory judgments. This indeed was the case in Germany.

In both Germany and France property rights and individual freedom are protected by law, but intangibles such as the right to a clean environment, health, or a certain quality of life are not actionable. Legal theorists have discussed the possibility of giving legal status to such rights. For example, some German writers advanced the case for a right to a decent environment, quoting the constitution that "each person has the right to life and physical entirety." [4] In France the former minister of justice, Jean Lecanuet, once proposed giving constitutional status to environmental rights.[5] But in both countries recognition of such rights has to come from the parliaments and cannot be established simply by broad judicial interpretation of property rights as in the United States. Despite the adherence of both French and German courts to the Roman legal tradition, attitudinal and structural factors shape actual practice. Thus the differences in the role of the courts in each country have been profound.

Available in both France and Germany are two avenues of challenge: the civil courts handle cases of personal damage and criminal action, and the administrative courts judge problems involving procedural breaches of the law, including in the case of Germany interpretation of the atom law.

Civil Courts

Under the civil code in both France and Germany nuclear energy is subject to strict liability, and the courts may compensate for personal injury or loss of property due to radiation.[6] In Germany the atom law of 1976 defines the damage awards and allows up to thirty years for claims. Utilities must be insured against accidents and the state is bound to meet additional claims up to a total of 1 billion DM. Theoretically environmentalists may also sue in the civil courts for radiation pollution. In Germany the maximum fine for exceeding the acceptable radiation level set by law is 100,000 DM per offense, but there have been no cases to date. In France the penalties are very low, ranging from 400 to 2,000 fr.[7] Following a scandal in 1975 over the dumping of radioactive effluents at two nuclear research centers, French environmental groups brought civil suit for pollution of the water table. A court-appointed committee found that the radiation level was 9 to 14 times greater than the maximum acceptable level allowed by law. The committee commented on the official silence surrounding the matter—the fact that routine dumping had been going on with the knowledge of a government oversight commission. But the court was reluctant to impose any penalty on the authorities.

Civil litigation often occurs in cases of expropriation of property. One of the very few cases that EDF lost in court was a clear violation of property rights when a farmer's fence was destroyed in the process of excavating a nuclear site.[8]

Illegal construction prior to obtaining a valid license has been another cause for legal action in France since the acceleration of the nuclear program in 1974. "Jumping the gun" is a common practice perceived as a way to avoid opposition. An internal EDF memo states this directly: "The best way to counteract local and national legal opposition is to get involved in the operation quickly and irreversibly and to make this decision public."[9] Environmentalists can bring such a case to a civil court under the penal code, but the fines are trivial. However, through a procedure called the *referé* or plea of urgency, the president of a civil court can make a provisional order to stop construction if he feels that harm is likely to occur.[10] Environmental associations brought three cases to court to block prepermit construction, but in each case courts ruled it unnecessary to submit the preparatory work to licensing requirements.

In Germany there have been fewer cases against illegal construction. KWU, as a private contractor, has less guarantee of government protection than Electricité de France.

The rules of standing—the right to sue—in a civil court are limited to those directly and individually affected. An environmental association may sue only to protect its own property or on behalf of the collective property interests of its members. Neither associations nor individuals can go to a civil court simply in the public interest.[11] In France the 1976 law on the protection of nature lifted some restrictions by allowing recognized as-

sociations to go before civil courts to uphold specific environmental statutes.[12] In Germany the concept of *Verbandsklage* or the right of an association to sue for the public interest has been debated for many years.[13] While
supported by the federal minister of the interior, a federal law conferring
such a right has never materialized.

Administrative Courts

The administrative courts were the major recourse for antinuclear groups.
The many permits and procedures required to license a power plant are all
subject to legal challenge, and environmentalists in both countries have
challenged every new permit in the administrative courts. The results of
these actions reflect basic differences in the organization of the French and
German judicial system.

The French system is centralized, consisting of twenty-four (twenty-five
including Paris) *tribunauxs administratifs* plus a national *Conseil d'Etat*. The *conseil* serves as a court of appeal. It is also directly responsible for cases affecting several regions or involving major administrative documents such as a
DUP decision. The centralization of this system has allowed close political
supervision by the ministry of justice. In contrast, the German system is
decentralized with a three-level hierarchy of administrative and appeals
courts. This structure allows dispersal of control through the *Länder* as well
as the federal authorities. Moreover the hierarchical system combined
with the multiple administrative procedures required to license a nuclear
power plant provide many points of appeal for aggrieved citizens. Activists
have effectively exploited these possibilities to delay nuclear power plant
construction.

The French system has more liberal rules of standing than Germany.
Registered associations and individuals with a personal interest, all those
near the area of a proposed plant, have standing to sue.[14] The French administrative courts have admitted actions even from nonregistered associations. In German administrative as well as civil courts, environmental
associations have standing only where the legally protected rights of the
association or its members are at stake. Yet in the case of nuclear power
this has been interpreted to include anyone living within a seven-mile
radius of the plant, and one judge granted standing to a man living sixty
miles away.

The ease of initiating legal action, however, bears little relation to its results. The consequences of legal action depend rather on the legislative context. The German judicial review has been based on interpretation of the atom law. Two decisions have guided judicial review. First, in 1972 the supreme administrative court ruled that of the stated goals of the atom law safety must take precedence over economic development and promotion of nuclear power. Second, the legislature in 1976 introduced an amendment to the atom law, providing that a nuclear facility may be licensed only after all safety precautions have been taken to the limit of science and technology.[15] Following these provisions, the German courts are able to consider substantive questions about both safety and technical feasibility (see table B.1 in appendix B). This of course often involves them in vague and difficult judgments. What level of risk is safe? What are the limits of current technology concerning possible safety precautions? The latitude inherent in dealing with such ambiguities has allowed some courts to take a very powerful and independent role in nuclear disputes.

France has no atomic energy law—only regulations and administrative decrees. These decrees differentiate nuclear plants from other large industrial facilities in that a separate operating permit is required; but this is not subject to full court review unless challenged on the grounds of illegal procedure (see chapter 3). EDF has not yet lost a major case.

French environmentalists routinely seek to suspend construction by arguing that there would be serious and irreparable consequences should work continue. Demands for a temporary injunction usually fail, but a revealing exception occurred following the implementation in January 1, 1978, of the 1976 law on the protection of nature. This law stipulated that an environmental impact statement was necessary for issuing a permit to build a nuclear plant and its absence was cause to grant a temporary work stoppage. On December 30, 1977, two days before the law came into effect, EDF obtained a construction permit for a plant at Flamanville where an earlier plea of urgency to stop prepermit construction had been turned down. An environmental association once again demanded suspension of work and to the surprise of all won its case. In the haste to acquire a permit before the new environmental law applied, EDF had tripped over a minor procedural technicality (see table B.2 in appendix B). However, EDF prepared a second application for a construction permit with the required environmental statement, and the project proceeded. Unless EDF violates a procedural rule, the courts will not force suspension of work. Once construction is underway, the possibilities for influence are virtually nil:

The entire process of building a nuclear reactor takes nearly ten years. The hierarchy of required authorizations should not create illusions. The only practical possibility to question the installation exists during the few months of the utility's application for a permit. However, this short period may be already too late because of the utility's important prepermit work on the site. Afterwards nothing more is possible.[16]

In Germany a law suit contesting a permit automatically requires suspension of work. But the licensing authority may by-pass this through an immediate effect order (IEO), insisting that its decision to issue a permit is in the public interest and therefore must take immediate effect.[17] Although the IEO is intended to be an exception rather than a rule, licensing authorities regularly use it as a means to prevent interminable judicial delays. Plaintiffs can still demand a temporary injunction or try in a preliminary hearing to lift the IEO and suspend construction. Their success rests on the judge's expectation about the eventual outcome of the administrative case, and his evaluation of the interests at stake.

To predict the outcome of a case, judges must assess if safety precautions have been adequate within the limits of available technology. Although the preliminary hearings are not intended to prejudice the later administrative proceedings, if a court does see fit to stop work on a plant, this can only reflect its judgment that the permit will ultimately be revoked for good reason. In this context judges have a great deal of discretion and power; in the Brokdorf decision, for example, the court lifted the IEO, arguing that, because of insufficient provision for final disposition of nuclear waste, the work should not proceed.

The latitude provided in the German legal system and the willingness of the courts to confront the government give the antinuclear movement an effective means of direct policy influence. But even in France antinuclear activists continue to take legal action, less for its direct effect than for its value in increasing public consciousness and augmenting the cost of nuclear power. In both countries then environmentalists have systematically initiated a deluge of law suits at every nuclear site.

Cases

French environmentalists have won few victories. Even if a permit is revoked on the grounds of a procedural violation as at Flamanville, EDF need only apply for a new permit. The major efforts to suspend construction in France challenged the legitimacy of the Declaration d'Utilité Publique (DUP) and the construction permit (AC) procedures (see chapter 3). In 1977 environmentalists lodged a complaint against the granting of a permit for the Super Phoenix fast-breeder reactor on the grounds that there had been no worthy public inquiry. The court rejected the complaint, arguing that a previous inquiry in 1974, although concerned with a preliminary trial project of quite different dimensions, had been adequate.[18]

In three cases environmentalists tried to suspend construction pleading urgency on the grounds that continued work would force an irreversible decision. Each time the civil court justices ruled themselves incompetent, claiming that the issue extended beyond their mandate to deal only with the legality of the permit and the adequacy of the procedures. In the first case, concerning the Super Phoenix in 1975, the courts claimed that EDF's preparation of the site was only preliminary, and no permit was necessary at this stage. Furthermore no questions of property rights were involved. Citizens made another plea of urgency at this site in 1977. A different judge responded essentially the same way, adding that most of the requirements for a permit had been completed. The third plea in 1977 at Flamanville also failed to stop construction (see table B.2 in appendix B).

These judges took advantage of the ambiguity in what actually constitutes construction of a nuclear installation. Yet refusal to question prepermit construction creates a paradox. As long as a civil court judge sustains EDF's contention that no work requiring a permit has actually begun, the case cannot go to an administrative court. As one environmentalist complained,

The administrative judges . . . can only annul a decision of the administration. . . . It is necessary that the decision exist and that it be known. . . . When EDF began work at Creys-Malville . . . and at Flamanville before having authorization, ecologists did not have access to an administrative judge. How, in effect, can one force a work stoppage when the decision authorizing the work did not exist or was not known? [19]

In Germany the responsibility to evaluate substantive questions of safety and the limits of available technology provide the courts with a wide margin of interpretation. The predilection of the courts thus becomes important, and since 1977 many judges have been increasingly critical of nuclear policy.

In an early case in 1973 at the Krümmel reactor the judge upheld the IEO requested by the licensing authority. The court reasoned that without evidence the permit would ultimately be revoked; the decision must rest on weighing the interests of the parties involved. Private interests lost out.[20] In a 1974 decision at Stade the courts followed the same reasoning.[21] In 1975, however, the Freiburg lower court ruling on Wyhl reversed the reasoning of the judges at Krümmel and Stade. It was the first decision to revoke an IEO. For this court the uncertainty about the outcome of later proceedings implied that the plant could not be deemed safe and that the health of the plaintiffs must come before economic interests. Furthermore, if work on this costly project continued, this could unfairly influence the final decision.[22] The Freiburg decision was overturned on appeal, and the lower court was instructed to stick to the law and avoid interpretations about the possible influence of a completed facility. In fact work was stopped, but as a result of the site occupation.

In 1977, as the main judicial proceeding on the Wyhl construction permit was underway, lower courts in Mülheim-Kärlich, Grohnde, Esensham, and Brokdorf all ordered suspension of the work. The judges in these cases were outspoken in their willingness to predict the outcome of administrative procedures concerning the permit and to specify substantive reasons why construction should cease. These included insufficient cooling capacity, insufficient detail on plans, and above all inadequate provisions for the disposal of nuclear wastes. By raising the waste disposal issue, the Brokdorf decision was a key point in the nuclear controversy. The reasoning behind this decision—that the atom law require provisions for the disposal of nuclear waste and that the problem had been inadequately considered—was to influence strongly future governmental as well as legal decisions. Later an administrative appeal court sustained the Brokdorf decision as a correct interpretation of the atom law and argued that the work stoppage should continue until specific provisions were made for long-term storage.[23]

In February 1977, when the courts suspended work at Brokdorf, environmentalists in Freiburg were contesting the Wyhl construction permit. Expert witnesses had been pronuclear, and the plaintiffs' arguments about the potential damage to the local economy and the dangers of low-level radiation were virtually ignored. Environmentalists predicted in mid-February that "the suit is as good as lost," and called the proceeding "irrelevant." One month later, on March 14, the court revoked the construction permit. Ignoring the concerns brought by the plaintiffs, the court decided that a worst-case scenario of an explosion within the reactor pressure vessel was potentially so catastrophic that it could not judge the reactor safe.[24] The Wyhl court strictly followed the reasoning of the atom law that safety must have precedence over economics and that precautions must be employed to the limit of the available technology. It added a third principle that had been stated by an appeal court in Münster in 1975—the greater the potential harm, the greater should be the precautions. It concluded that despite the small chance of disaster, and the high cost of technological improvements, greater safety measures were necessary. It was the plant, not the safety, that must be scrapped if public health is at stake. In the words of the court,

Should the necessary protection remain technically impossible, then the precautions required by article 7 [of the atom law] would not be guaranteed. . . . Since the protection of life and health is at stake, precautions cannot be limited simply to presently feasible safeguards.[25]

The Wyhl court by no means set a precedent that was rigorously followed. Just two weeks after its decision the Würzburg administrative court used the same principle to uphold the construction permit for the Grafenrheinfeld nuclear plant. This court explicitly rejected the idea that a worst-case risk was grounds to revoke a permit and ruled that the plant was safe within the bounds of existing technology which was all the atom law required. It also rejected the Brokdorf reasoning, denying that long-term questions such as waste disposal were within the scope of the court's discretion.[26]

Legal Competence and Political Constraints

The courts in France have avoided extending their competence beyond procedural considerations, and despite the flood of legal actions they have remained peripheral to the nuclear debate. The German courts, confronting the central substantive question of nuclear safety, have been thrust to the forefront of nuclear politics. The differences reflect a set of political relationships that have turned out to be crucial for nuclear policy.

Given the legal situation, the French courts are unable to confront the administration on substantive grounds and for the most part have served only to prevent gross procedural irregularities. The civil courts declined to make use of the tools available under the plea of urgency procedure to halt prepermit construction of nuclear installations, and the administrative courts avoided decisions that would embarrass the administration.

Three structural factors prevent independent legal action in France: the absence of a legal statute that clearly lays out the principles of safety for nuclear power plants, the limited role of judicial review that would allow substantive considerations, and the existence of EDF as a powerful bureaucracy with monolithic control over nuclear policy. One critic calls this situation a "codification of our trust in the technicians of the atom."[27]

The 1976 law on the protection of nature requiring environmental impact statements has not yet significantly improved the legal position of environmental associations. They have long had access to the courts on procedural matters; their problem is rather how to obtain a favorable decision on substance. In 1977 the FFSPN initiated a series of legal actions against no less than twenty-nine reactors under construction because they had no environmental statements.[28] Filed in early 1977 before the environmental law came into effect, this effort was intended more to influence public opinion than to gain a substantive victory. For the role of the French courts in the implementation of environmental law remains attenuated by their reluctance to deal with more than the technicalities of procedure. An environmental lawyer, however, sees possibilities in the new legislation, claiming that laws allowing more room for judicial interpretation combined with the recruitment of a new generation of judges could change both administrative and judicial attitudes:

The laws for environmental protection are new. To punish a polluter is not yet fully accepted by public opinion. But this can quickly change. Look at the field of work accidents: at one time factory owners were never punished. When several directors of large firms were put in jail because of negligent safety precautions in their factories, it was a shock for "the good society," and some newspapers campaigned against "the young generation of red judges." But suddenly in the mid-1960s public opinion woke up to the importance of the issue. Who knows if such things will happen in the field of environmental protection.[29]

In Germany, where the courts can already rule on substantive issues of safety, a different situation prevails. Yet the German courts are not inherently more open to the arguments of political opposition than their French counterparts. In fact traditionally the bar in Germany has been conservative, linked closely to the power structure of the state. Judges were mostly from established families. Initially reluctant to challenge the administration directly, they avoided questioning public facilities such as nuclear power plants. As late as 1976 a legal scholar observed that the courts rarely exercised their power of judicial review to probe administrative decisions. "Especially in atomic energy cases no decision is ever decided in favor of the plaintiffs."[30]

Given this tradition, some observers attribute the recent judgments on nuclear safety to the caution of a conservative institution. Basing judgment on literal interpretation of the law could be seen as a conservative and legalistic approach. But this explanation neglects the recent expansion of personnel in the administrative court hierarchy. When the social-liberal coalition in the government extended activities in such areas as urban planning, environment, and work safety, this also extended the hearings and administrative appeals. Today over one-quarter of all German judges are involved with disputes between the individual and the administration. Many of these judges were recruited in the early 1970s from a generation of young graduates influenced by the student movement and its goals for a more democratic society. Willing to defy traditional attitudes in so far as possible within the bounds of the law, they are less concerned with defending the general interest than protecting the citizen from the state. A lawyer describes a prevailing attitude among these young judges:

One advantage in presiding over an administrative rather than a civil court is that we don't have to contradict our personal convictions by punishing individuals. Another is that we have a real chance to act according to our political philosophy by curtailing further uncontrolled bureaucratic power and preventing further restrictions on human rights and freedoms. Judges with a left ideology will usually try to find a job in the administrative rather than the civil courts.[31]

The atom law in Germany provided an opportunity for these judges to exercise their predilections in interpreting the safety of nuclear power. They have generally avoided ruling on such imponderables as the local economic and environmental effects of power plant construction. Instead the atom law, reinforced by the administrative supreme court's ruling that safety takes priority over economic considerations, allowed them to deal with the fundamental issue of safety and the question of whether the utilities had taken adequate precautions to the limits of existing technology. Perhaps most important in explaining the recent court decisions is the decentralization of the German political and administrative system—a structure that precluded the kind of monolithic pressure that restricted the judgments of their French colleagues.

The response of politically established groups to the increasingly aggressive position of the German courts is decidedly mixed.[32] Some praise the courts for their courage, for giving the country a breathing space, a bit of time to reconsider the issue of safety before it is too late. Others charge the courts with stepping beyond their sphere of responsibility. Politicians were understandably distressed, in some cases arguing that the courts lack the political competence to deal with the issue and in other cases that they lack adequate technical competence. A union leader wondered whether "decisions about jobs in West Germany should be left up to four or five judges."[33] The CDU/CSU responded to the Wyhl ruling by calling for an amendment of the atom law to prevent further politicization of the judiciary. The federal minister of justice, Bernhard Vogel, proposed an amendment that would categorize nuclear plants by type, requiring permits only for each type rather than for each plant. Such standardization would reduce the discretion of the courts.

Even the courts had doubts about their apparently expanded scope and power in this area. An appeals court in Münster, ruling on the construction permit for the breeder reactor prototype at Kalkar, suggested that decisions concerning reactors are too important for the courts and better made in parliament.[34] In August 1977 it deferred its decision pending a ruling by the federal constitutional court on the constitutionality of the atom law. Should parliament assume direct responsibility for nuclear licensing, then the only avenue of appeal would be the constitutional court, and the only issue open to appeal would be the constitutionality of parliamentary decisions. This would essentially remove the courts as an avenue for direct public influence.

Other proposals for structural change followed the critical court decisions. The presidents of ten appellate courts suggested that administrative proceedings be limited to a two-tier system in which cases would go directly to the appellate level, by-passing the lower courts. The administrative supreme court would then be the only court of appeal. Another proposal would limit the interpretive power of the courts by specifying future energy needs in the atom law, thereby removing this from judicial deliberations.[35] Such proposals for limiting the role of the courts flourished as rulings placed a de facto moratorium over Germany's nuclear program.

The role of the courts has been the decisive factor in determining the outcome of the nuclear debate. In Germany the courts served as a useful political resource for antinuclear groups, and the possibilities of effective legal action helped to sustain the movement. That such possibilities for influence were unavailable to French activists contributed to their decline as an effective political force. Moreover the role of the courts extended beyond the specific siting decisions considered; for anticipation of legal action also served to modulate the nature of the political response to the antinuclear movement.

12 The Government Response

Government officials were not prepared for the intensity of opposition to nuclear power. To many the movement appeared as an attack against the state, science and technology, modernism, and certainly the public interest. They had to respond—and not only to technical questions about risk. For the antinuclear movement challenged the very legitimacy of the decision-making process.

The response included a variety of information programs intended to win public consent, repressive measures to intimidate opponents, and institutional or procedural changes to co-opt critical groups. The character of these activities in France and Germany reflected traditional attitudes toward dissent and differences in political organization. In France, where dissent and demonstration have been a ubiquitous feature of the political culture, protest is often dismissed more as a resistance to change, a defense of special interests, or a result of public ignorance than as a major political challenge. Indeed the French establishment, in the manner of a grand bourgeois accustomed to challenge but assured by experience that he will not lose his position, tends to ignore protest as long as possible.

In Germany the antinuclear movement raised questions of legality, legitimacy, and constitutional principle. Officials often attribute opposition to communist infiltration or anarchism. Protest raises anxieties that radical dissent could destroy social order, and consensus becomes a crucial objective.

Underlying these attitudes toward dissent lie very different patterns of national continuity. French territorial unity has existed for at least six centuries, providing a basis for a centralized and stable state. German unity was never defined by territory but rather by the mythological heritage of the Roman Empire. Until Bismarck founded the *Reich* in 1870, Germany as a state had no political continuity. While the concept of nation in France was never directed outside the territory, referring only to national sovereignty, the German concept has stressed cultural, linguistic, and ethnic characteristics, distinguishing between a *Kulturnation*, comprising all who speak the same language and share the same culture, and a *Staatsnation*, those living in a given territory. In the past the hope of unifying the German state, of realizing the identity of the culture and the nation, pervaded German ideology. It was only after World War II that national consciousness in Germany adopted a concept of nation that coincided with its territorial boundaries.

Ambiguity in the definition of nation and state in Germany left a fear for national stability that has been reflected in the exclusion of those who criticize the basic social order. In the nineteenth century the Bismarck *Reich* defined its identity by stigmatizing the political center and the social democrats as "enemies of the *Reich*." [1] The chaotic decline of the Weimar Republic, the war, and the division of the German nation in 1949 reinforced insecurity and the fear of dissent: the radical left has been systematically excluded from political life in the Federal Republic since 1956 when the constitutional court prohibited the Communist party.

A popular analysis of contemporary Germany explains this low tolerance for conflict and dissent in psychoanalytic terms. It attributes Germany's emphasis on public order to an inability to face past reality, an inability to mourn.[2] Whatever the explanation, the emphasis on consensus has created on the one extreme a widely accepted delegation of power to oligarchic elite and on the other a persistent radical criticism, rejecting the very existence of the state and expressing its disaffection in subversive terms. Critical groups such as antinuclear activists often become identified with this tendency.

The contrast with France is remarkable. French national identity is never questioned: there is solid social consensus on what it is to be a member of the French nation. The authority of the centralized state was in jeopardy only for brief revolutionary periods in French history—during the French Revolution and the Paris Commune—but in both cases the principle of centralization was quickly reinforced. While protest appears endemic in France, national continuity has reduced sensitivity to dissent. Despite rapid political change and the many different constitutional regimes since the beginning of the century, only the Vichy period experienced significant repression of civil liberties. Yet there persists in France a pervasive belief in the need for strong central authority, capable of containing the centrifugal forces that could threaten the capacity of the state to serve the general interest.[3]

The political composition of the governments in the two countries also affects their response to protest. The Gaullists and later the Giscardists have identified with the state and tend to ignore social movements as long as they do not significantly influence elections. In Germany the social-democratic-liberal coalition government is committed to seek broad public support for its policies. In addition the federal government cannot ignore the independent positions taken by the *Länder* governments. These features influence the ways in which the governments of France and Germany seek to contain protest.

The Contradictory Effect of Information

The French government, regarding nuclear policy as a technical implementation of its energy plan, left the burden of public information to EDF. Until 1975 the utility distributed information on nuclear power as part of its effort to promote energy consumption; it dismissed opposition as the voice of obscure marginal groups hostile to progress. An EDF advertisement in 1974, for example, pictured an attractive bikini-clad woman next to a caveman: "Can anyone argue that the human species has degenerated?" The ad explains that cavemen lived with 125 millirems of natural radiation, but their average lifespan was only fifteen years: Today with the same level of natural radiation plus artificial radiation we live to seventy years.

The utility brought local officials, members of parliament, and the press to model power plant sites—usually Saint Laurent-des-Eaux, with its three heated swimming pools, parks, and evidence of prosperity. It published a five-page cartoon strip on the benefits of nuclear power for children. A set of pamphlets emphasized on the local tax and employment benefits of nuclear power, essentially dismissing potential environmental or social problems related to the technology.

By 1975 EDF found itself engaged in a veritable battle of brochures. When the utility distributed 35,000 copies of a brochure called "Fessenheim: Source of Prosperity for Alsace," antinuclear groups responded with "Fessenheim: Life or Death for Alsace." When EDF prepared a technical dossier for high schools, GSIEN presented its own version of the technology.

The government then appointed a delegation to coordinate information on nuclear power and help regional councils participate in the consultation procedures required to site power plants. In April 1975 this delegation distributed 100,000 copies of a book providing economic, technical, and environmental information on nuclear energy. EDF also formed a nuclear information group and opened a documentation center in Paris, making available technical information on reactor safety.

As the promoter of nuclear power EDF had problems of credibility, giving the ecologists considerable advantage in gaining public confidence among educated middle-class groups (see table 12.1). Thus, as opposition persisted, EDF tried to withdraw from public disputes. Refusing to debate with environmentalists at Braud St. Louis, an EDF official argued that a public discussion was not warranted: "Municipal officials and deputies already exist; an adversary procedure would allow neither useful dialogue nor objective debate." [4]

When forced to respond following the demonstration in Creys-Malville and the electoral success of the ecologists in municipal elections, the French government initiated a systematic information program. It was assumed that more information would reduce public fear.

In 1977, a few days after Creys-Malville, President Valery Giscard d'Estaing announced the creation of the Conseil d'Information sur l'Énergie Electro-nucléaire (Council for Information on Nuclear Energy). The group did not meet for some eight months; the government feared that public discussion of nuclear energy could reflect unfavorably on its campaign for the March 1978 election. Finally, in April 1978 the eighteen members of the council, including mayors of several municipalities housing nuclear power plants, environmentalists, physicians, and journalists, convened. Mme. Simone Weil, then minister of health and one of the most popular members of government, was its first president. The council reviews the information on nuclear power available to the public, evaluates

Table 12.1
Credibility of sources of information

	Readers of Express[a]	SOFRES survey[b]
Ecology activists	49%	14%
Journalists	29%	6%
EDF direction	18%	30%
Industry	8%	6%
Government spokesmen	7%	17%
Mayors	4%	7%
Members of parliament	3%	2%
No opinion	7%	7%

Source: *Express* (April 19–26, 1975).
Note: The question asked was, Who do you trust most in questions of nuclear energy? Informants could make several choices.
[a] 25,000 readers of *Express,* mainly educated middle-class, responded.
[b] The sample size was 1,000.

its quality and completeness, and on the basis of scientific consensus rec-
ommends what should be publicized. It can neither generate its own information nor force material to be released. During its first months of work
the council convinced the prime minister's office to release reports on radiation exposure prepared by the government safety agency (Service Central de Protection contre les Rayonnements Ionisants). But it failed to obtain controversial emergency evacuation plans until after the Three Mile
Island accident when authorities agreed that they should be gradually
released.

In its evaluations the council must assume there is consensus among
scientific elites; its function is mainly to discover this consensus to judge
what should be presented as fact to the public. Those who strongly diverge
from this view feel excluded from hearings. Environmentalists in the council, frustrated with its limited role, often threaten to resign.[5]

While the French information effort took the form of a council, the German program was a citizens' dialogue. Far more participatory and based
more on adversary principles than expertise, the German information program reflected a very different perception of the role of the public in decision making.

The coalition parties, SPD and FDP, were acutely sensitive to the nuclear protest and its possible political consequences. Fears were reinforced
by the experience in Sweden where the social-democratic government of
forty-four years had failed over the nuclear issue. Thus, when the government began to accelerate the nuclear program in 1974, it also organized a
massive public information program. This citizens' dialogue was to provide information on the position of the federal government, to promote
political and technical discussion, and to support opinion-forming activities within adult education groups, churches, labor unions, parties, and
other associations.[6]

The government organized its own public discussions and supported debate organized by parties, labor unions, the churches, and adult education
institutions. It insisted that in all meetings both pro- and antinuclear viewpoints be exposed. It provided considerable resources for this dialogue:
750,000 DM in 1975, 3,000,000 DM in 1976, and 4,000,000 DM in 1977.[7]
The minister for science and technology, Hans Matthöfer, opened the procedure by announcing the official opinion in the press and publishing
800,000 copies of three handbooks providing information on technical
questions of safety and on the economic and social dimensions of nuclear
policy as well.[8] He emphasized that, while the dialogue could open new

perspectives, the government remained convinced of the need to expand nuclear power.[9]

The official government position dominated the early dialogue, and citizen initiatives dismissed it as a pronuclear propaganda campaign. But the structure of the program encouraged public debate, and the antinuclear initiatives skillfully and effectively used the dialogue to publicize their own viewpoint and increase their own influence and credibility. Indeed as the dialogue progressed, the opinions of the citizen initiatives, tied neither to government nor to industry, appeared relatively unbiased. A German opinion poll in 1977 showed that these initiatives were among the more' credible sources of information on environmental and nuclear issues (see table 12.2).

It is difficult to assess the effect of the citizens' dialogue on public attitudes toward nuclear power. It exposed thousands of people to the controversy and might well have reinforced skepticism, especially among the members of youth organizations. It certainly gave political leaders some sense of public opinion. But it clearly failed to create a public consensus in support of nuclear power. One objective of the dialogue, however, was to dispel criticism that decisions about nuclear energy were made in an undemocratic way: "The confidence of the people in the functioning of democratic process and in the justness of the political decisions which have been made in this field must be reinforced or—where lost—must be restored." [10]

Table 12.2
Credibility of sources of information on environmental problems

	INFAS survey
Scientists	64%
Medical doctors	37%
Citizen initiatives	26%
Municipalities	15%
Journalists	11%
Politicians	10%
Utilities	6%
Industry	6%
Others, no answer	11%

Source: INFAS, 1977. The sample size was 1,196.
Note: The question asked was, If you receive information on environmental problems created by nuclear power, who would you rather trust? Informants could make several choices.

The contrast between the nuclear information policies in France and Germany reflects different concepts of democracy: the French elitist information model corresponds to a style of policy making that derives democratic legitimation from electoral majorities; the German participatory model corresponds to its cooperative policy-making process, in which democratic legitimation proceeds through a consensus among major social forces. The French procedures left no doubt about the government's intention to continue with its nuclear program. In Germany the government insisted that new options could come out of the citizens' dialogue if it could be proven that conservation and other developments can cover energy consumption.[11] But the German effort faced the classic dilemma of a liberal government seeking to act as if central planning can proceed without the use of power--as if every policy can be based on consensus, as if all parties to a debate can somehow come to terms.[12] While suggesting the open availability of options, the German government also clearly intended to implement the nuclear program, a paradox that appeared to critics as hypocrisy and contributed to the dissent.

Intimidation

Public scrutiny and accountability limit the use of force to contain dissent. Whenever police entered to break up antinuclear demonstrations, public attitudes polarized and government credibility suffered. A French opinion poll right after the Creys-Malville demonstration found that only 52 percent of respondents agreed with the government's decision to prevent the site occupation at any price, and 40 percent supported the ecologists' plan to maintain the demonstration despite official prohibition.[13]

To avoid the loss of legitimacy that follows the use of force, governments seek benign means of control. Antinuclear groups report cases of intimidation in the form of research restrictions, employment threats, or personal harassment. In France the unions offer some protection. For example, when Louis Puiseux, a high official in the economic-planning division of EDF, published *Babel nucléaire,* a book critical of nuclear energy, rumors circulated that he was asked to resign.[14] His union, CFDT, reacted vigorously, calling the incident a witch hunt. EDF's director promptly denied having taken any action against Puiseux, and it was suspected that the initiative came rather from the ministry of industry who hoped to prevent protest within the state utility.[15]

In another case an ad hoc research group that included people from

EDF and CEA did a government-financed sociological study on the nuclear debate that contained certain points perceived as critical of EDF's and CEA's policies. Some officials in these organizations attacked the research report as biased and lacking scientific quality. CEA's public relations director asked that the leader of the research be fired but to avoid a conflict similar to the Puiseux incident, this step was not taken. But the research team had to reorient its work toward technical risk assessment studies.

In Germany where the antinuclear movement is often associated with radical anticonstitutional behavior the legal possibilities of harassment are greater than in France. According to an agreement in 1972 among the ministers of the interior of the German *Länder* (confirmed by the constitutional court in 1975), candidates for civil service positions (15 percent of the employed labor force) must guarantee that they will actively stand for the liberal-democratic constitutional order. This agreement, the *Radikalenerlass,* has led to routine inquiries on the political past of candidates. Newspapers and magazines have reported numerous cases where the *Verfassungsschutz* investigated antinuclear activists and labeled them anticonstitutional.[16]

The *Radikalenerlass* has opened many possibilities for intimidation; for example, the government of Hamburg prohibited teachers from wearing the antinuclear button, "Nuclear Energy? No Thanks!" It was argued that wearing such a button during work was agitation and political propaganda in the classroom, in violation of the constitutional obligations of civil servants. The German labor unions have not reacted to protect their members who are critical of nuclear policy, and some have even taken action against antinuclear activists within their ranks. Thus the possibility of effective individual intimidation is greater than in France.

The problems of intimidation were dramatized by the highly publicized harassment of an official in a nuclear firm. Klaus Traube was one of three top managers of the German nuclear development firm Interatom. Suspected of association with the terrorist milieu, Traube was watched by the *Verfassungsschutz,* and his mail and telephone calls were monitored. When these observations failed to confirm suspicions, the *Verfassungsschutz,* with permission from the ministry of the interior, illegally installed hidden microphones in Traube's home. Again no evidence of conspiratorial relations with terrorists emerged. Nevertheless the *Verfassungsschutz* pressed Traube's employer, Siemens, to fire him. Siemens at first refused: "Only three persons exist worldwide with sufficient competence and qualifications to fill this post: among them Dr. Klaus Traube." Traube was at that time responsible for the fast-breeder project in Kalkar, and the police feared that his access to sensitive material could be exploited to build weapons.[17] Interatom denied this as a technical possibility, but Klaus Traube was eventually fired.

Der Spiegel discovered and published the story as a serious violation of civil liberties. The police had constructed a case from few facts and many prejudices: Traube had been at a party with one of the defense attorneys for members of the Baader-Meinhof group. He had belonged to the Communist party at age 17, and his mother had been a communist decades before. These facts and his "unusual" private life (Traube is divorced) were used against him. The scandal had no immediate political effects, but to nuclear critics it illustrated their fears of a coming atom state.

Intimidation is hardly sufficient to contain a mass movement. Authorities in both France and Germany sought means to discourage participation in demonstrations and divide the antinuclear groups. They distinguished serious environmentalists from the subversive elements which according to government officials infiltrated the movement. The prefect at Creys-Malville labeled German demonstrators as violent anarchists interested in disturbing public order. He tried to mobilize local hostility against the traditional enemy: "Our villages are once again occupied by the Germans!" Before the demonstration in Hannover over the Gorleben nuclear waste facility, the minister of interior of Lower Saxony announced that he had prepared sufficient space in prisons for violent demonstrators.[18]

Verbal association is but one form of collective intimidation. The mayors of the villages close to Creys-Malville prohibited camping to discourage people from coming. Local officials imposed heavy fines for disturbing public order or invading private property to prevent site occupations. In June 1979 in violation of international agreements French authorities stopped several thousand Germans suspected of trying to join a demonstration from crossing the border into France. Every car with an ecology sticker was stopped at the frontier, and the German police cooperated by giving the names of 600 antinuclear activists to the French police.[19]

Such tactics, however, also turn public opinion against the government. The unfavorable public reaction to the violence at Brokdorf constrained the official response at subsequent demonstrations. In Grohnde the local government simply ignored the protest by giving it no publicity at all. While the Brokdorf demonstrations had been preceded by an extensive press campaign, Grohnde was hardly mentioned in the press until after the demonstration. Then at Kalkar the government chose another tactic: it gave wide advance publicity as the demonstration was organized but avoided an aggressive tone; the government as a friend and servant would do everything to guarantee a peaceful demonstration. Then identities were checked and vehicles searched throughout the region.

The violent clashes of 1977 raised difficult questions: Is the government willing to take measures that would turn the right to demonstrate into a farce? How much coercion is a state willing to use to implement an unpopular policy? Just as violence had forced the ecologists to reconsider their tactics, so it forced government officials to realize the limits of intimidation as a means of social control.

Co-optation and Delegitimation

To mobilize political support for its policies, the German government encouraged the formation of pronuclear citizen initiatives. In 1978 secretary of state for environmental protection in the federal ministry of the interior, Gunter Hartkopf, proposed that industry should create its own citizen groups. Industrial representatives replied that this indeed was their intention.[20] In fact some 20,000 people belong to pronuclear citizen initiatives, created and financed by nuclear firms.

Among the most important of these groups is the Aktionsgemeinschaft der Bürgerinitiativen für Energiesicherung und Kerntechnik, AEK (Action Committee of Citizen Initiatives for Energy and Nuclear Technology), a national organization based in Alzenau, Bavaria. Its leaders are from utilities, nuclear, and construction firms. This committee collaborates with Atomforum, using the same information sources and strategic advice from leading marketing firms. It primarily seeks to mobilize support from influential citizens but also copies BBU tactics, distributing pronuclear bumper stickers, buttons, and leaflets.

Labor union activists in more than five hundred firms involved in energy production belong to the Aktionskreis Energie der Betriebsräte (Action Committee Energy of Labor Representatives). This was the committee that organized the demonstrations in Dortmund before the 1977 SPD party congress. It had captured union representatives from major nuclear, metallurgy, and construction firms, but this strategy was politically effective only so long as the committee's industrial ties were not visible. When industrial manipulation became apparent after the SPD congress in 1977, the DGB publicly dissociated itself from the committee and criticized it for creating a pernicious faction within the union.[21]

The German government also tried to integrate the antinuclear and environmental movements into the conventional political process. In 1971 it formed the Arbeitsgemeinschaft für Umweltfragen (AFU), a consortium of one hundred fifty organizations including environmental associations, ministries, public authorities, and private firms, to promote public awareness of environmental problems and to propose and review legislation. AFU now receives more than half of the government subsidies for all environmental associations. The association holds an annual Environmental Forum on such topics as Cooperation between State and Industry, Environmental Protection and Economic Policy, Environmental Protection and Siting—An International Problem, and Citizen Initiatives—Citizen Participation.[22] One of its four permanent working groups is on nuclear energy and radiation. The AFU represents an effort to coordinate environmental groups and bring them into the policy process. However, effective integration requires compromise, and tensions within the environmental movement prevent AFU from becoming a significant and effective lobby: Representatives from traditional associations are willing to negotiate with industry and government, but the more political antinuclear groups refuse to compromise on the nuclear issue. They do, however, maintain contact with AFU, finding it a useful source of information and a welcome opportunity for the personal encounters necessary to dispel negative stereotypes.

In France the effort was less to co-opt the antinuclear movement than to discredit it. In 1972 the minister of the environment sponsored monthly lunches with environmentalists to discuss environmental policy. These meetings generated the Comité de la Charte pour la Nature, formed to draft a charter for nature. The charter, drafted in November 1972 by twenty traditional associations, proclaimed that the protection of nature was the very basis of civilization and made a number of recommendations concerning budget allocations to rural areas and the formation of advisory environmental councils with representation from the associations. Inspired by environmentalists with active political responsibilities in the majority parties, the charter's recommendations were sufficiently general to allow agreement among environmentalists and sufficiently diplomatic to allow negotiation with the administration. Indeed the administration considered the document a valuable contribution to its environmental policy. But significantly the charter made no mention of nuclear power, and, when one of its scientist-politicians, Louis Le Prince-Ringuet, defended the nuclear program, the Comité lost much of its credibility in the eyes of environmental groups.

Other official efforts to deal with environmental problems also ignore the nuclear issue. In 1979 the parliament created Comité Législatif d'Information Ecologique (COLINE). This group of ten parliamentarians, eight representatives from environmental associations, and eight legal and scientific experts serves as a bridge between elected officials and associations to integrate the associations into political decisions and assure a better implementation of environmental laws. All political groups except the Communist party are represented. Again COLINE's initial documents avoided the issue of nuclear power.

The strategies to contain opposition to nuclear power depend on political traditions. In Germany the bargaining process that takes place prior to policy decisions encourages efforts to support those social forces in agreement with the government and organize groups that could one day enter the social partnership.

In France the government is relatively independent from social-bargaining procedures and hardly interested in developing structures that would further complicate its work. It rather seeks to turn people away from the more politicized ecology groups and minimize the political salience of the nuclear issue.

Procedural Innovation

Despite government attempts to contain the protest against nuclear power and co-opt opposition groups, skepticism persists. Increasing disagreement among scientists exacerbated the dilemma. How are regulatory authorities and legislators to make judgments when faced not only with diverse political interests but also with conflicting evidence and polarized expert opinion? How can decisions be made about the acceptability of risk given the confusion among facts, their interpretation, and their subjective evaluation? Attention increasingly turned to the nature of decision-making procedures—an issue that had from the beginning been central to the antinuclear movement.

Reforms of decision-making procedures followed from an assumption that appropriate procedures to evaluate risk would reduce conflict and lead to public acceptance. We have already described reforms of the li-

censing process, the French DUP, the new environmental impact statement legislation, and the regulations of the German atom law. In France
these procedural changes tend to be relatively nonparticipatory, rather
emphasizing the importance of expertise. New laws that give environmental associations a voice in administrative and legal procedures are restrictive, allowing little possibility for effective influence.

The relationship established between the new ministry of the environment and the environmental associations is a case in point. A special division, Service d'Information et de Relation avec les Associations de
l'Environnement (SIRAE), exists to inform officially recognized associations about ministerial activities and new procedures such as environmental impact statements and industrial-licensing requirements. The point is
to enable associations to use this information in their own activities and to
negotiate more effectively. But SIRAE and other consultation mechanisms
organized through the new ministry include only those major associations
officially recognized and thereby exclude the more critical groups. Moreover they are intended less as a partnership than a means to create a constituency for the minister of the environment as he confronts the administration with environmental demands.

The German decentralized decision-making context has provided ecologists with greater political opportunity, because they can play one administration against another. In a politically controversial area such as
nuclear policy, each administrative and political level tends to delay decisions and shift responsibility. Each political body tries to maintain an
image of liberalism and openness by increasing public consultation. The
result is a dilution of responsibility and authority; indeed reform proposals
intended to clarify responsibilities for nuclear policy are usually aborted.
For example, after Wyhl the government proposed to shift responsibility
for siting and licensing power plants from the administration to parliament. This would have removed the possibility for citizens to challenge
policies through the administrative courts. The proposed reform, however,
played into the hands of the ecologists by reinforcing their contention that
the nuclear issue encouraged undemocratic administrative behavior. This
in itself reduced the proposal's political feasibility and precluded its
implementation.

To maintain a liberal image, local politicians tried to increase review
and consultation procedures. They accepted the increased time necessary
to reach a decision as the price of participation and expected that demonstrations of goodwill would ultimately win support for government policy.

But tactical openness had unexpected and crucial consequences. The Gorleben International Review was a striking example.

In 1976, when the government in Bonn asked Ernst Albrecht, prime minister of Lower Saxony, to choose a site for the combined reprocessing and waste disposal complex, antinuclear groups immediately attacked the plan and the *Land* government agreed to organize an independent international expert review. As a member of the pronuclear CDU in a *Land* where the SPD was skeptical of the nuclear policy, Albrecht needed to emphasize both his openness on the issue and independence from the Bonn government. Addressing the *Land* parliament, Albrecht declared, "I can guarantee you one thing: this complex will never be realized if the political situation continues. The SPD on the federal level demands its construction, the *Land* government shows its face and gets beaten, and the SPD on the *Land* level continues its polemics against the complex." [23] Albrecht hoped that the international review would gain time and help to reconstruct political consensus.

The expert panel was to review some 3,000 pages of the technical reports that had concluded the project was feasible and would cause neither environmental damage nor excessive health risk. In April 1978 the *Land* government hired Helmut Hirsch, an Austrian physicist and administrator to coordinate the Gorleben International Review. His experience as organizer of the Austrian nuclear information campaign and his status as a foreigner was expected to guarantee a competent and unbiased approach. With advice from the leaders of citizen initiatives Hirsch proposed a panel of twenty experts, most of whom were known for their critical attitudes toward nuclear power.[24] The government accepted this idea.

At their first meeting in Hannover in September 1978 the panel charged the Lower Saxony government with obstructing the review and insisted on more adequate information and translations of the safety reports into English. The key information was translated and the panel prepared its assessment of the project.

The Gorleben International Review ended in early April 1979 with a debate chaired by the physicist Carl Friedrich von Weizsäcker, known for his pronuclear position. For six days the international panel of scientists confronted experts from the nuclear industry. Prime Minister Albrecht, many government officials, and the press attended; at stake was the largest industrial complex in Germany. The press portrayed antinuclear experts as well prepared and scientists from the nuclear industry as unable to answer some of their key questions.[25] The Three Mile Island accident, which occurred during the debate, gave further weight to the antinuclear arguments; it also helped activists mobilize one of the largest demonstration to date.

Evidence of increasingly hostile public attitudes turned the SPD in the *Land* against the project, at which point Prime Minister Albrecht had to make a decision. Declaring support in principle for nuclear energy, he turned down the proposed project: "We failed to convince the public of its necessity and safety."

The Gorleben International Review was explicitly driven by traditional party politics: it was the local political situation more than the substance of the dispute that forced a technical reassessment of the project. This suggests that extraparliamentary pressure can have considerable influence if an issue becomes a crucial point of negotiation within the establishment.

The decision to abandon the combined reprocessing and waste disposal plant and concentrate on an intermediate storage facility could stop further nuclear energy expansion in the Federal Republic, for construction of power plants is linked by law to the solution of the waste disposal problem. At the very least the opposition to nuclear power in Germany has paralyzed the official decision-making process and caused long delay. Chancellor Schmidt has declared his intention to proceed with the nuclear program regardless of opposition, but as a journalist put it, "While it is too soon to write an obituary for the death of nuclear power, criticism has become respectable." [26]

The contrast in the government response to the antinuclear movement in France and Germany reveals some striking ironies. The French government uses less intimidation in its response to radical forms of opposition but has few procedures through which critics can obtain information or influence decisions. The German government is more willing to enter into negotiation with ecologists and has a more open information policy but reacts more vigorously to radical dissent. However, in both countries the extensive experimentation with informational and procedural innovations largely failed to integrate the antinuclear movement within the framework of traditional political life. There are several reasons. First, most of the innovation did not respond to the basic sources of conflict. While the authorities, especially in France, focused almost exclusively on questions of risk, antinuclear activists saw these risks as a surrogate for larger social, political, and economic concerns. Indeed, even if technical consensus could have been established, this would have had little effect on antinuclear attitudes.

Second, institutional and procedural innovations were largely dependent on information provided by the government, the industry, or the utilities. Their effectiveness in reducing conflict necessarily required credibility and trust. But it is precisely a lack of trust and credibility that drives the opposition to nuclear power.

Third, effective negotiation procedures work only when the protagonists share a minimum common interest that will lead to mutually reasonable compromise. Government efforts to integrate groups into a negotiating framework had little effect because the opposing groups did not share the values that would compel a compromise.

Finally, the government response failed to squelch the opposition because most procedures involved structured discussions over predetermined policy. The financial and administrative investments involved in nuclear energy were so great that politicians and administrators had only limited maneuverability.

High technology creates a basic political dilemma. Government dependence on industry for technological development inevitably reduces political credibility. Caught between constraints imposed by alliance with industrial interests and the need to maintain an image of liberal-democratic behavior, the governments could only oscillate between declarations of good will on the one extreme and repression of protest to maintain order on the other.

The greater range of tactical oscillation in Germany is largely due to the decentralized structure and the interpenetration between government and multiple social forces. Government agencies and ministries in Germany tend to be tied to specific constituents; yet they must also integrate new social movements into policy negotiations. Contradictions and intra-governmental conflicts are inevitable—between, for example, regional and national governments or between environmental agencies and the ministry of economics. The antinuclear movement increased its own influence by exacerbating such intra-administrative conflicts. The resulting inconsistencies in government policy reinforced rather than reduced public mistrust.

In France, in contrast, a centralized administration allows the government to ignore social movements. In the end it was less the government's response to the antinuclear movement than its power to implement policy despite continued protest that allowed France to continue her nuclear program while Germany's is indefinitely delayed.

V INTERPRETATIONS

13 The State of European Nuclear Programs

At least six years have passed since European nations planned major increases in their nuclear programs—years that have been marked by intense opposition in nearly every European country. While each country has its unique program for nuclear power development, their policies are closely interrelated. France and Germany, together with the United States, are the main nuclear plant suppliers for most of the smaller European nations.[1] Thus decisions about nuclear power in these smaller countries seriously affect the market for the French and German nuclear industry. Conversely any significant slowdown of the programs in these two countries would greatly increase skepticism abroad. The antinuclear movement itself extends beyond national borders. Demonstrations attract activists from many countries, and the ecology press is sufficiently coordinated that the actions in any one country are closely followed elsewhere. This chapter briefly describes the different affects of the nuclear debate on the evolution of nuclear policy in the smaller European countries. These examples extend our analysis of the political and institutional factors that shape national policies toward controversial technologies.

In late 1978 after years of antinuclear demonstrations and an extensive government information campaign the Austrian people turned down their nuclear energy program in a national referendum.[2] The vote in Austria was close but definitive—a result that came as a profound shock to the socialist government, which along with industry and the labor unions was committed to the technology. A nuclear power plant costing over $500 million—a considerable investment for a small nation—was all but ready to go into operation. After the decision this project was abandoned.

In February 1979 a similar referendum took place in Switzerland. Nuclear critics who had been active in Switzerland since 1975 had demanded that responsibility to approve the siting of nuclear plants be transferred from the central authorities to the people living within a 20-mile radius of the proposed site. This was the referendum issue, and it was narrowly defeated when 51 percent of the voters supported continued Federal Government authority, thereby allowing continuation of the nuclear program. Conscious of the strength of opposition, however, the Swiss government proposed to tighten its licensing procedures by requiring utilities to provide evidence of need when ordering a nuclear plant and guarantees of safe disposal of nuclear waste. In addition it proposed parliamentary approval of the licensing procedure. Although this theoretically would remove power from the federal bureaucracy, environmentalists criticized the proposal as a tactic to create a false sense of security while allowing the program to continue with what in fact would be minimal constraint. Voters, however, approved these government proposals in a second referendum.

In Sweden public opposition to nuclear policy had been the primary cause of the 1976 electoral failure of the social-democratic government after forty-four years in power. Until 1970 Sweden's nuclear policy had developed with the consensus of the five main parties and the important interest groups; a pattern of consensus maintained by compromise prevailed.[3] Thus, when grassroots protest first emerged in Sweden in the mid-1960s, it was ignored. The growth of the environmental movement, however, encouraged the Center party to change its policies and to take an antinuclear position in 1972, just at a time when the government was planning a major nuclear expansion. By breaking the establishment consensus, the Center party provided an influential and official channel for popular dissent. Subsequently in 1976 increasing differences among the political parties over nuclear power turned this issue into an electoral theme. The new center-right coalition government, strongly influenced by the antinuclear Center party, initially imposed a de facto moratorium on all new power plant construction. But only several weeks after coming to power the government approved a controversial plant at Barseback, a heavily populated area near the Danish border. This became a major target for antinuclear groups, and demonstrations involving Danish and Swedish environmentalists took place throughout 1977. By late 1978 divided over the future of nuclear power and unable to come to a decision about future growth of the technology, the center-right coalition split. Public opinion about nuclear power meanwhile relaxed: polls indicated that those against nuclear power declined from 55 percent in 1976 to 37 percent in 1978, and those in favor increased from 28 to 41 percent. Then the Three Mile Island accident once again hardened public opinion against the technology: after the event 53 percent of those polled were against nuclear power and only 26 percent were still favorable. Anxious to keep nuclear policy out of the 1979 parliamentary elections, the Social Democrats called for a special referenda in March 1980. The referenda included three alternatives each backed by political organizations. The choice of the Center party and forty-five antinuclear groups was no new construction and a phase-out of existing reactors within ten years. It won 38.6 percent of the 6.3 million votes. The Social Democrat/Liberal party choice of six new reactors, but with state and municipal control, and a phase-out in twenty-five years received 39.4 percent. The Conservative party and industry choice of six reactors with no state control won 18.7 percent. While all parties agreed to abide by the referendum, antinuclear leaders convinced of a broad and committed base of public support expect to continue resistance.

In Holland strong opposition came from the Labor party, the media, public action groups, and many scientists. Antinuclear groups included many prominent people from within the establishment. The combined pressure from these groups indefinitely delayed plans for the construction of three plants that were intended to provide 20 percent of Holland's electrical-generating capacity. Dutch activists have participated in the "Stop Kalkar" movement and in efforts to block the construction of the centrifuge uranium enrichment plant at Almelo, owned by a German-British-Dutch consortium.

In Italy political debate, site demonstrations, and the growth of citizen groups have forced substantial delay in the government's plans. In Spain several protests against the nuclear program for the Basque coast in 1977 and 1978 attracted over 100,000 demonstrators.

What factors have shaped the development of nuclear energy programs in these nations? Among those factors what has been the effect of the antinuclear movement on government policies? Obviously the degree of energy dependence is of major importance: those countries relatively rich in primary energy resources, such as the Netherlands with its natural gas, or the North Sea countries with their oil, planned relatively small nuclear energy programs and were receptive to public concerns. Countries poor in resources, in general those in southern Europe, organized major nuclear programs and are understandably reluctant to cut back their initial plans.

While the availability of resources and short-term fluctuations in energy consumption have clearly had important influence on nuclear power expansion, nuclear power decisions are also mediated by several political factors: the structural relationships that form between governments and their nuclear industries and the available channels through which critics are able to exercise influence over public policy decisions.

Public ownership characterizes the energy sector in all European countries. We have suggested that even in Germany where there are private utilities public authorities have a decisive stake in the promotion of nuclear power through shareholding. Furthermore this state involvement in the energy sector influences the actions of other social and political organizations—parties and unions—as they seek to maintain a consensus on major and costly public investments. Government and industry necessarily became partners in the costly enterprise of nuclear power. The very cost of nuclear policy—the size of past investment in the nuclear option—tends to reduce government autonomy in this policy area and limit its ability to respond to public concerns.

The pressure of nuclear plant suppliers on government policy is especially important in those countries with national nuclear firms. France, Germany, and Sweden each have their own national nuclear industries, and their governments have taken direct economic responsibility for this industrial sector through supporting research and development and by maintaining long-term contracts with nuclear plant supply firms. Nuclear policy in these three countries is a sector where the theory of state monopoly capitalism clearly applies. The high capital costs of developing a national nuclear program brought close collaboration between these governments and their industries even before the decisions were made to commercialize nuclear technology on a large scale. Collaboration grew rapidly after 1974, when the sharp price increase of oil brought pressure to expand alternative energy sources, and these three governments soon became the most active promoters of nuclear energy at home and abroad.

By 1974 several of the smaller countries, Belgium, the Netherlands, and Italy, were already linked by shareholding or contracts to one of the major multinational nuclear consortiums. As internal criticism of nuclear policy developed, only a few countries, Austria, Denmark, and Norway, still had the option to renounce this technological choice.[4] In these cases the lack of commitment to a national nuclear industry and the relatively limited economic ties between government and nuclear plant supply firms allowed governments to maintain greater regulatory independence and responsiveness to public concerns.

However, the ability of opposition groups to influence policy rests only in part on this degree of commitment. More important, as we have seen in France and Germany, are certain political and administrative arrangements: the official channels for conflict resolution (such as the court system), the possibilities for citizens to participate in administrative decisions at different government levels, and the openness of large political organizations to dissident popular demands.

The state of the nuclear energy debate in different European countries has reflected such differences in administrative structures and political traditions. In Austria the vote against nuclear energy was the result of a national referendum organized by the government. This was a risky strategy, only possible in a country without a national nuclear industry and with a government basically in control of the energy sector. Private corporations in Austria had relatively little power to put decisive pressure on the government, which was therefore able to respond to public demands. The referendum in Switzerland, with its constitutional elements of direct democracy, was forced on the political agenda by an activist citizen initiative.

In Holland, Germany, and Sweden the existence of official channels through which the public could influence substantive policy decisions gave the opposition the means to delay the expansion of nuclear power. Holland was in the throes of widespread administrative reform directed toward democratization of policy decisions in many sectors. Experiments in citizen involvement were underway in the area of regional planning during the late 1960s. In the planning of major technological programs, elaborate procedures had been developed to incorporate public feedback at the earliest stages of policy making. In this pluralist society conflict and accommodation are accepted as a political reality, and there are many points of entry for dissenting groups.

In Germany the ecologists had created a political climate that was relatively sympathetic to their cause. Then backed by the atom law, and its provisions concerning the disposal of nuclear waste, they were able to stop several projects through the courts. In addition the dispersal of power in a federal system provided the possibility of influencing policy through the *Land* governments.

In most countries, including both France and Germany, the political parties played essentially no role in the nuclear debate. Sweden is a striking exception. It is one of the few countries where the decision to slow down the nuclear program took place within the traditional political process. Here years of conflict among the political parties over the proposed nuclear program essentially paralyzed the decision-making process.

Political and legal structures in Germany, Austria, Holland, and Sweden provided channels for public influence and points of tension through which activists could create divisions within the political establishment. In France the lack of such institutions allowed the government to continue its nuclear program essentially unchanged. Public protest was as widespread and intense in France as in other countries; however, neither the courts, the parties, nor the local governments could channel protest to the policy level.

The continued controversy over nuclear power throughout Europe has had critical economic ramifications. In response to the 1973 oil crisis, the European commission had developed a long-term energy plan for the EEC countries. The plan would have decreased dependence on foreign raw materials from 63 to 40 percent by reducing the percentage of oil, maintaining internal coal production (which had been declining for some fifteen years), and increasing the share of nuclear energy in electricity production. The EEC expected to see 172 plants in operation by 1985. According to an internal EEC study in 1976, however, this rate of growth was far too rapid relative to energy needs. The contracts the utilities had signed with nuclear suppliers would result in overproduction of electricity by 1979; costs had been grossly underestimated, and a nuclear park planned for 1985 would force the closing of conventional power plants long before their obsolescence.[5]

But then by 1979, examining the actual implementation of national nuclear programs, the EEC revised its nuclear forecasts downward. In 1973 it had projected there would be 160,000 Mw in production by 1985; in 1979 it expected only 71,000 Mw. The psychological and political impact of the Three Mile Island accident was predicted to delay further most nuclear programs (except in France) for over a year.

Limited storage and reprocessing facilities in Europe have contributed to delay. Meanwhile the economic benefits of nuclear power relative to other sources of energy are increasingly in question. New safety measures, compelled by the Three Mile Island accident, have compounded economic problems. In France and Germany the problems of the nuclear industry are further complicated by the limited expansion of the Third World markets, due to political instability and the growing concern about nuclear proliferation.

Does this mean that one can look forward to an end of the controversy over nuclear power? Even without further construction of new plants those countries that have already implemented a nuclear program will have to live with the long-term technical problems of waste disposal and the decommissioning of older facilities. The existing plants and the waste already accumulated will force solutions that are likely to face continued opposition. The nuclear industry has created these technical problems, and it now must solve them to survive as an economic entity. It will also have to live with serious political problems that have been exacerbated by controversy.

The association of government with the unpopular nuclear industry has been very costly in terms of public trust. In those countries committed to nuclear programs basic questions are raised about the efficacy of regulation. Can the state act at once as the promoter and regulator of a technology? Does nationalization provide better control in the interest of the public or does it simply reinforce industrial goals? Can a system be designed to provide real public control over utilities and supplier firms? Just a few weeks after the Three Mile Island accident, German Chancellor Helmut Schmidt reminded nuclear experts at a European nuclear conference that the industry owed the public reliable information about risks. He warned that, unless public concerns were taken seriously, the loss of confidence could have profound influence on the future of nuclear power.[6] In the early days of nuclear power development critics were forced to prove their claims against the official image of the technology as a safe and clean source of energy. Schmidt's warning suggests that the burden of proof has shifted to the promoters of nuclear energy; the nuclear establishment must convince the public that the problems concerning the safety of power plants and the critical elements of the fuel cycle (reprocessing and waste disposal) are technically under control. This is perhaps the major policy significance of the nuclear controversy.

14 Social and Political Significance

The controversy over nuclear power has had direct influence on energy policy. But what is its social and political significance? By challenging a major government program, nuclear critics raised difficult political questions about how decisions are made in advanced industrial sectors. Does this imply a possibility for change in the political landscape? Certain patterns in the origin and the evolution of the antinuclear movement in France and Germany help to guide our interpretation.

Origins of Extraparliamentary Dissent

The costs, the scale, and the complexity of the nuclear power program requires administrative structures characterized by economic concentration, close cooperation between industry and the state, and increasing use of expertise as a basis of decisions. These properties are hardly unique to the nuclear power sector. For years all established political groups have favored such administrative arrangements as a way to implement widely shared commitments to rapid industrial growth and technological change.[1] However, in the early 1970s a growing environmental movement called rather for control of technology and limits to growth. The structures that had been organized to generate growth and promote technology became increasingly suspect. These concerns have guided the evolution of the antinuclear movement and its sociopolitical analysis.

Nuclear critics see the social and political consequences of nuclear power through very different lenses than the promoters of this technology. Their vision diverges on such varied issues as energy dependence, safety, and civil liberties. Conflicting perceptions prevail about the role of government and the appropriate use of scientific expertise in the decision-making process. These differences, summarized in table 14.1, have been a source of profound political strain. Yet tensions were seldom expressed within the traditional political institutions. The heavy capital investment in nuclear technology created a sense of economic responsibility and commitment among political leaders that closed off political debate. Democratic institutions such as parties, parliaments, and unions had only a limited role in the articulation and mediation of conflict in this technological arena. Ten-

Table 14.1
Conflicting perceptions in the nuclear debate

Antinuclear analysis	Pronuclear analysis
Political consequences	
Government and industry are in collusion with little reference to broader political goals.	Government and industry only serve to implement agreed-upon political objectives.
Nuclear power implies dangerous concentration of political power and an omnipotent bureaucracy.	Government acts in the public interest. Bureaucracy is necessary for efficiency.
Nuclear power encourages proliferation and can lead to war.	Availability of energy reduces international tension.
Economic and social consequences	
Nuclear power reinforces dependence on American technology.	Nuclear power reinforces national independence.
Nuclear power means economic concentration and further inequities.	Nuclear power is necessary for growth and full employment.
Nuclear power implies a police state that threatens civil liberties.	It is the protest and the threat of terrorism that threatens civil liberties.
Role of government	
Government should defend small units against large concentrations.	Government should defend public interest against special interests.
Government should protect future generations against harm from today's generation of energy (nuclear waste).	Government should assure that future generations have adequate resources by conserving fossil fuels.
Role of scientific expertise	
Science can be manipulated for alternate ends.	Science is neutral.
Science can be a source of harm as well as benefit.	Science contributes to progress.
The problem is one of the acceptability of risk. This limits the value of technical evidence.	Technical evidence is the only basis on which to evaluate risk.

sions were rather expressed outside of official political channels through forms of extraparliamentary dissent.

This pattern of conflict illustrates three closely related political developments in industrial democracies: the crisis of legitimacy, the politicization of society, and the questioning of expert counsel in the formulation of public policy.

Considerable discussion has focused on the so-called crisis of legitimacy, the growing disaffection with democratic institutions unable to fulfill their traditional role as mediators of social conflict.[2] Crucial decisions about controversial policy issues are increasingly generated outside the representative political process, evolving less from compromise among competing interests than from consensus among expert elites. However, depoliticized modes of planning do not imply the end of politics. They have rather stimulated another syndrome that has been called the politicization of society.

The antinuclear groups are but part of a larger social movement comprised of citizen initiatives and action groups seeking direct influence over specific policy issues affecting their interests.[3] Some of these activist groups are defined in terms of their goals (such as environmental protection); others organize in terms of a collective identity (ethnic, regional, or women's organizations). As such groups struggle to influence policies through direct political action, they also challenge the forms of technocratic politics that have evolved at the highest political levels.

Thus the third development illustrated by the nuclear power controversy is the questioning of expert counsel and its role in the formulation of public policy.[4] Scientific expertise serves as a source of legitimacy and consensus—as an authoritative and compelling standard of rationality and objectivity in the decision-making process. In the nuclear debate the governments of both France and Germany tried to restrict the controversy to its technical aspects, hoping that a consensus among experts would finally emerge to solve the social conflict. But, as the antinuclear movement established its own relationships with counterexpertise, factional disputes developed within the scientific community. The very fact that scientists disagreed about the safety of nuclear power exacerbated political conflict, undermining the use of scientific evidence as a consensual basis of political choice. The different experiments to win public acceptance of new technologies—the expert reviews, inquiries, and information programs (chapter 12)—further enhanced the role of experts as a political resource. For as governments sought to balance competing scientific views, expertise became less a source of consensus than a weapon in the political arsenal of different social groups.

Reflecting these three political developments, the antinuclear movement gained a diverse constituency and elaborated an ideology and a vision of social change. Its significance, however, has been subject to diverse interpretations.

A Harbinger for Social and Political Change?

The antinuclear organizations in France and Germany share a number of structural features. Diffuse and decentralized, they include diverse groups with different political agendas. The most active ecologists are young, well educated, and often technically trained. In both countries, however, support for the movement extends well beyond these activist groups to blue-collar workers concerned about safety and to farmers who often participate in siting disputes.

The movement is coordinated by a committed leadership, an active ecology press, and a discourse marked by powerful apocalyptic and political themes. This discourse perpetuates the values that had shaped the student movement of the late 1960s. But the themes of the movement also include classical issues of industrial conflict and technological change: safety, health, working conditions, land expropriation, and labor substitution. They also embody preindustrial concerns recalling the late nineteenth-century revolt against expanding industrialization.

Despite such similarities antinuclear groups in Germany and France developed in ways that reflected the significant differences in their political environment. The role of traditional voluntary associations, the structure of the political institutions confronted by the protest groups, and the way the governments responded to dissent shaped the tactics of this social movement and determined its ability to influence nuclear policy. In Germany the multiple administrative levels provided many more points of articulation for protest groups than the centralized structure in France. The existence in Germany of an atom law subject to interpretation by the courts was an especially powerful resource. As governments sought to manage dissent by administrative reforms, they further increased the channels through which tension could be expressed. The greater impact of the protest in Germany than in France followed less from the characteristics of the movement itself—a movement often paralyzed by internal schisms over tactics and long-term goals—than on these points of access within the political system and on the ability of activists to exploit them.

Expectations about the possibility of influencing nuclear policy accounts for differences in the mobilizing capacity and indeed in the survival of the movement in each country. In Germany after a latent period following the violence of the 1977 site occupations, the antinuclear movement once again mobilized in April 1979 to demonstrate against the Gorleben project. Six months later in October 1979 antinuclear demonstrations in Bonn drew some 150,000 people. Ecology groups have felt sufficiently confident to present a green party in the general elections of 1980.

In France we see a very different picture. The definitive electoral victory of the presidential majority in 1978 destroyed the ecologists' hope for more favorable conditions under a left government, leaving the movement in a slump, unable even to mobilize after the Three Mile Island accident. The economic crisis affecting France more than Germany further reinforced the differences in the movement's vitality in these two countries.[5]

Given such differences, what is the long-term significance of the antinuclear movement? Some observers, the most prominent of them the sociologist Alain Touraine, base their interpretation on the pervasiveness of antitechnocratic themes in many recent social movements. They point to the growing contradiction between the citizen and the technocratic state, to the increasing disaffection with democratic institutions unable to fill their traditional role as mediators of social conflict, and to the proliferation of action groups seeking direct policy influence. They predict that these struggles will become more important than the classic antagonism between capital and labor and that the new social movements will revitalize the organized left as a means of social and political transformation.[6]

Our study, however, suggests that this is unlikely in Germany or France, for ecologists disaffected with the traditional left seek above all to maintain political autonomy. Moreover the antistate and antiauthoritarian dimensions of their ideology contradict the tradition of disciplined collective action so important to the labor movement. In fact the gap between ecologists and the organized left remains profound.

Others interpret the antinuclear movement as a remedial force, a corrective to prevailing technological trends:

The growth of the ecology movement and its demonstrations over the past ten years is not just the passing fashion among a privileged and irresponsible youth; it is a sign of health, a production of antibodies against the cancer of technology uncontrolled, a healthy reversal of the myth of indefinite material growth. The antinuclear movement may not be the avant garde of social progress . . . but the concentration of the anxieties of our time on nuclear power is a sign of a fear that, if we leave technology to the power establishment, our advanced society will be shattered.[7]

Such an interpretation is valid, however, only if a social movement can actually exercise some influence, if the political process provides opportunities for substantive intervention. The very survival of a movement as a remedial force depends on its opportunities for substantive intervention. This becomes apparent in comparing the French and German movements. In Germany nuclear critics have in fact affected government policy through effective lobbying and skillful use of the courts. Furthermore it appears that the electoral challenges of the ecologists may have a decisive influence on the government majority and in the long-run significantly alter the German three-party system. Even a small percentage gain in general elections could shift the balance of power from the social-liberal coalition to the Christian-democratic opposition, and this could eliminate the Liberal party from parliament and eventually from political life. This very prospect of electoral influence is a source of political leverage. The governmental parties have not been able to come to grips with their internal conflicts over nuclear power, and the German government has had to adopt an extremely cautious strategy in proceeding with its nuclear program. "In our democratic system based on public participation," claims Chancellor Schmidt, "nuclear energy cannot be implemented without broad public consent."[8]

The possibilities of influence in France are quite another story. The French government continues to dismiss criticism; in the situation of increasing unemployment, fear of economic crisis, and a fragmented left new social movements have lost much of their impetus, and their substantive impact on nuclear policy remains limited.

The major significance of the antinuclear movement may lie in its expression of widespread resistance to certain trends in advanced industrial societies. The antinuclear discourse conveys a diversity of social concerns and a fermentation of political ideas. The apocalyptic criticism, the global views, and the sense of crisis evoked by the prospect of a nuclear society appear more as a form of utopian idealism than a coherent political vision. But the movement reflects concerns that extend far beyond the issue of nuclear power. It suggests both an ideological resistance to the reductionism epitomized by science and technology and a political resistance to the bureaucratization of the social system. The themes of the movement express a challenge to the legitimacy of the political order and a willingness to oppose authority through extralegal action. The seductive vision of an ecological society based on aesthetic values and holistic ideals expresses the powerful desire for a new world view, a cosmology, a set of absolute ethical standards that would resolve the pervasive sense of social crisis.

This search for an all-embracing ideology could lead to a kind of ecological fundamentalism, turning some away from political involvement and leading others toward acts of political despair. The movement could thus encourage the formation of alienated factions and extremist sects. But more often the ideas of social movements become absorbed in the culture, shaping future political patterns and cultural beliefs. Indeed it is likely that after a period of discomforting cleavage and conflict the utopian image of social harmony embodied in the ecology discourse may become a means of repressing continued social contradictions.

Questions of the significance of a social movement must necessarily be speculative, but whatever its long-term destiny the antinuclear movement has posed major challenges to the predominant scheme of interpretation in contemporary society, raising questions about the future of technological development and the social values, priorities, and political relations that underlie technological decisions.

Appendix A: Nuclear Power Plant Sites

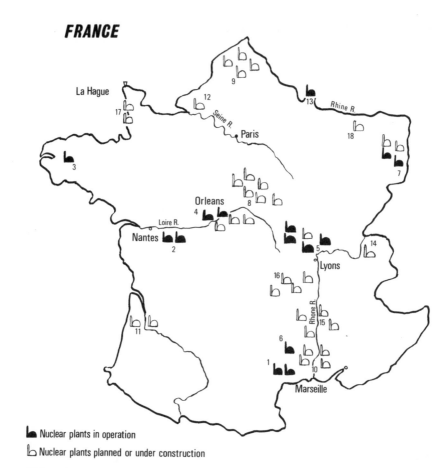

Figure A.1
Nuclear power plants in operation, under construction, or planned in France.

Table A.1
Nuclear power plants in operation, under construction, or planned in France (1979)

In operation	Mw
Marcoule G2 (1)	40
Marcoule G3	40
Chinon 2 (2)	210
Chinon 3	480
Monts d'Arée (3)	70
Saint Laurent-des-Eaux 1 (4)	480
Saint Laurent-des-Eaux 2	515
Bugey 1 (5)	540
Bugey 2	925
Bugey 3	925
Bugey 4	905
Bugey 5	905
Phoenix (6)	233
Fessenheim 1 (7)	890
Fessenheim 2	890
Fessenheim 3	905
Total	8,953

Under construction or planned by 1985	Mw
Saint Laurent-des-Eaux 3 (4)	925
Saint Laurent-des-Eaux 4	925
Saint Laurent B1	905
Saint Laurent B2	905
Fessenheim 4 (7)	905
Dampierre 1 (8)	925
Dampierre 2	925
Dampierre 3	925
Dampierre 4	925
Gravelines B1 (9)	925
Gravelines B2	925
Gravelines B3	925
Gravelines B4	925
Tricastin 1 (10)	925
Tricastin 2	925
Tricastin 3	925
Tricastin 4	925
Blayais 1 (11)	905
Blayais 2	905
Blayais 3	905
Blayais 4	905
Chinon B1 (2)	925
Chinon B2	925
Palvel 1 (12)	1,300
Palvel 2	1,300
Palvel 3	1,300
Palvel 4	1,300
SENA (France/Belgium) (13)	280
Super Phoenix (14)	1,200
Cruas 1 (15)	905
Cruas 2	905
Cruas 3	905
Cruas 4	905
Saint Alban 1 (16)	1,300
Saint Alban 2	1,300
Saint Alban 3	1,300
Saint Alban 4	1,300
Flamanville 1 (17)	1,300
Flamanville 2	1,300
Cattenom 1 (18)	1,300
Cattenom 2	1,300
Total	41,835

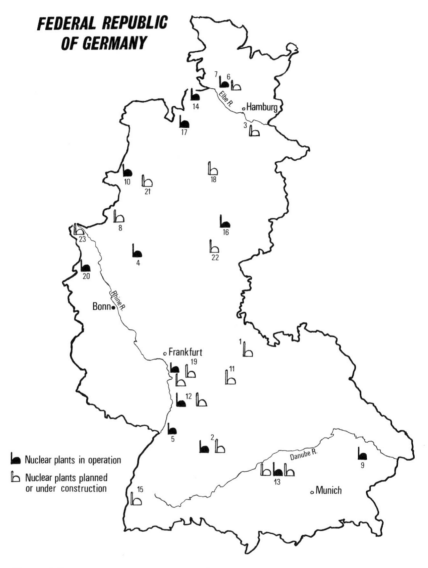

Figure A.2
Nuclear power plants in operation, under construction, or planned in West Germany

Table A.2
Nuclear power plants in operation, under construction, or planned in West
Germany (1979)

In operation	Mw
Neckarwestheim I (2)	856
Hamm (4)	300
Karlsruhe (5)	58
Brunsbuettel (7)	805
Isar (8)	907
Lingen (10)	250
Obrigheim (11)	345
Philippsburg 1 (12)	900
Stade (14)	662
Würgassen (16)	670
Esensham (17)	1,300
Biblis A (10)	1,200
Biblis B	1,300
Total	9,553

Under construction or planned by 1985	Mw
Grafenrheinfeld (1)	1,300
Neckarwestheim II (2)	845
Krümmel (3)	1,316
Hamm (8)	1,300
Philippsburg 2 (12)	1,365
Gundremmingen B (13)	1,310
Gundremmingen C	1,310
Biblis C (19)	1,303
Biblis D	1,303
Borken (22)	1,300
Vahnum I (23)	1,300
Vahnum II	1,300
Total	15,552

Stopped by courts	
Brokdorf (6)	1,365
Wyhl (15)	1,362
Grohnde (17)	1,316
Kalkar (20)	327

Appendix B: Legal Actions

Table B.1
Examples of German court cases

Nuclear plant	Permit contested	Plaintiffs	Court	Decision	Reasoning
Suspending effect cases (motions to revoke IEO and/or to impose a temporary injunction)					
Krümmel 1973	1972 provisional site approval with IEO	Weltbund zum Schutz des Lebens and 4 neighbors	VG Schleswig Holstein	Motion rejected. Appeal confirmed lower court decision.	Permit did not effect the declared purpose of the plaintiff's environmental association. Expert opinion does not anticipate that permit would be revoked. Public interest must override individual concerns.
Stade 1974	1972 seventh partial construction permit with IEO	An individual	VG Schleswig Holstein	Motion dismissed. Appeal confirmed lower court decision.	Claim inadmissible at late stage in permit proceedings. Public interest greater than personal plaintiff's safety concerns.
Wyhl 1975	First partial construction permit	Four communities	VG Freiburg	IEO revoked. Work suspended. Appeal overturns lower court and work continues.	Outcome could not be determined, therefore plant not implicitly safe. Life of plaintiffs comes before economic interests. Presence of completed facility would influence final decision. Appeal denies right of lower court to interpret effect of completed plant on later proceedings.
Mühlheim/Kärlich 1977	Construction permit	An individual	VG Koblenz	Temporary injunction. Work suspended. Appeal overturns lower court. Construction continues.	Plans had changed since original approval. Collusion between utilities and authorities.
Grohnde 1977	Partial construction permit with IEO	An individual	VG Hannover	IEO revoked. Work suspended.	

Esensham 1977	Operating permit	An individual	VG Oldenburg	IEO revoked. Work and operation stopped pending main proceeding.	Insufficient cooling capacity.
Brokdorf 1977	First partial construction permit		VG Schleswig Holstein	IEO revoked. Work suspended. Appeal confirmed lower court.	Inadequate consideration of disposal of nuclear waste. Appeal referred to atom law's requirement for nuclear waste disposal plans and required application for immediate storage facility and geological testing for final waste disposal facility.

Administrative court actions

Würgassen 1968	Construction permit		VG Minden. Appeal in OVG Münster	Permit upheld. Appeal confirmed lower court.	Cannot discuss effects of long-term, low-level radiation during a construction permit procedure.
Würgassen 1972	Operating license		Supreme administrative court.	Withheld pending legal clarifications. Now in operation.	Debate over safety and the availability of expert advice to the courts.
Wyhl 1977	First partial construction permit		VG Freiburg. Appeal in OVG Mannheim	Permit revoked. Appeal underway.	Insufficient provision for containment of reactor pressure vessel. (Main issues lodged by plaintiffs—dangers to crops, heat pollution of Rhine, radiation pollution—were not considered.) Issue was priority of safety over economic interests.

Table B.1 (continued)

Nuclear plant	Permit contested	Plaintiffs	Court	Decision	Reasoning
Grafenrheinfeld	Construction permit		VG Würzburg	Permit upheld.	Atom law does not proscribe limits to radiation exposure. Competence of court does not extend to nuclear waste disposal. Containment of reactor vessel not necessary.
Kalkar 1977	Continued lawsuits since 1973		VG Münster	No decision.	Must wait until constitutional court rules on constitutionality of atom law. Eventually decided that parliament must decide on breeder reactor. Parliament approved construction.

Note: VG = administrative court; OVG = appeal court

Table B.2
Examples of French court cases

Nuclear plant	Permit contested	Plaintiffs	Court	Decision	Reasoning
Suspending effect cases (motions for a *surcis à execution*)					
Super Phenix Breeder 1977	DUP and DAC	Conseil Général de l'Isère	Conseil d'Etat	Motion dismissed.	Early inquiry in 1974 was inadequate.
Flamanville 1978	DUP, DAC, construction permit, and permission to build a platform in public waters.	CRILAN and group of farmers	Tribunal de Caen. Appeal in Conseil d'Etat.	Motion granted. Appeal confirms lower court.	Authorization to build platform delivered prior to construction permit. Technicality contrary to normal procedure and construction would prejudice later decision.
Plea of urgency					
Super Phenix 1975		Mouvement Ecologique Rhône-Alpes and l'Association pour la Sauvegarde du Site de Bugey et Malville	Tribunal de grande instance de Bourgoin-Jallien	Rejected	Judges declared themselves incompetent to deal with issue. Work was preliminary and therefore no permit required and no expropriation involved.
Super Phenix 1977		Association pour la Sauvegarde du Site de Bugey et Malville and FRAPNA	Tribunal de grande instance de Lyon	Rejected	Most of requirements for DAC were completed. Claim of right to quality of life dismissed as nonexistent
Flamanville 1977		CRIPAN CRILAN	Tribunal de grande instance de Cherbourg	Rejected	Most of requirements 1 DAC were completed. of right to quality of lil dismissed as nonexisten

Notes

Chapter 1

1. For an analysis of this concept of technology, see Langdon Winner, "Do Artifacts Have Politics," *Daedalus* (winter 1980), pp. 121–136.

2. *International Encyclopedia of the Social Sciences,* vol. 14 (New York: Macmillan and Free Press, 1968), pp. 438 ff.

3. See A. Touraine, *La Voix et le regard* (Paris: Le Seuil, 1978).

4. O. Rammstedt, *Soziale Bewegung* (Frankfurt: Suhrkamp, 1978), pp. 173 ff.

Chapter 2

1. R. Jungk, *Der Atomstaat* (München: Kindler, 1977).

2. A. Gorz (M. Bosquet), *Ecologie et politique* (Paris: Le Seuil, 1978), pp. 114 ff.

3. For the early days of French nuclear policy, see L. Scheinman, *Atomic Energy Policy in France under the Fourth Republic* (Princeton: Princeton University Press, 1965); J. C. Bupp and J. Derian, *Light Water: How the Nuclear Dream Dissolved* (New York: Basic Books, 1978).

4. The different backgrounds of professions in CEA and EDF reinforced tensions between the organizations. While both institutions recruit their personnel from the graduates of Ecole Polytechnique, most of the engineers of CEA come from the specialized school Ecole des Mines and EDF is dominated by graduates from the Ecole des Ponts et Chaussées.

5. See L. Puiseux, *La Babel nucléaire* (Paris: Editions Galilée, 1977), pp. 112–123.

6. *Europa Yearbook 1977: A World Survey* (London: Europa Publications, 1977), p. 615.

7. F. de Gravelaine and S. O'Dy, *L'Etat EDF* (Paris: Alain Moreau, 1978), pp. 269–270.

8. For detailed information on this network see P. Allard, M. Beaud, B. Bellon, A. M. Lévy, and S. Lienart, *Dictionnaire des groupes industriels et financiers en France* (Paris: Le Seuil, 1978), pp. 136 ff.

9. See OEEC-Armand Report: L. Armand, *Some Aspects of the European Energy Problem* (Paris: OEEC, 1955). K. Winnacker and K. Wirtz, *Das unverstandene Wunder. Kernenergie in Deutschland* (Düsseldorf: Econ, 1975), pp. 338 and 153.

10. M. Meyer-Renschhausen, *Energiepolitik in der BRD von 1950 bis heute* (Cologne: Pahl-Rugenstein, 1977), pp. 136 ff.

11. C. Deubner, *Die Atompolitik der westdeutschen Industrie und die Gründung von Euratom* (Frankfurt: Campus, 1977), pp. 14 ff.

12. For more information see *Deutsches Atomforum* (Bonn: Tätigkeitsbericht, 1977). For a detailed analysis see L. Mez, "Die Atomindustrie in Westeuropa," in *Technologie und Politik,* 7 (1977), pp. 146 ff.

13. For example, the *Land* North Rhine-Westphalia is an important shareholder

in the RWE, the *Land* Baden-Württemberg in the Badenwerke AG, and the *Land* Bavaria in the Bayernwerke AG.

14. De Gravelaine and O'Dy, *L'Etat EDF.*, p. 142; L. Mez, *Die Atomindustrie*, p. 161.

15. These are mostly the three largest German banks: Deutsche Bank AG, Dresdner Bank AG, and Commerzbank AG.

16. Interview with a CEA official.

17. D. Schnapper, *Morphologie de la haute administration française* (Paris: Mouton, 1969).

18. Winnacker and Wirtz, *Das unverstandene*, p. 78.

19. Ibid., p. 352.

20. A government report lists the professional origins of 927 experts of the BMFT: researchers and teachers, 510; representatives of industry, 232; representatives of other interest groups, 46; others, 139. *Bericht der Kommission für wirtschaftlichen und sozialen Wandel* (Bonn: 1976), p. 250.

21. For a sociological analysis of techniques for harmonizing conflicting interests, see L. Boltanski, "L'Espace positionnel. Multiplicité des positions institutionnelles et habitus de classe," in *Revue française de sociologie*, 14 (1973), pp. 3–26; M. Pollak, "L'Efficacité par l'ambiguité," in *Sociologie et sociétés* (spring 1975), pp. 29–49.

22. The following professional profiles illustrate how high officials circulate between different sectors: Pierre Guillaumat, administrator of the CEA (1951–1958), president of the Petroleum Research Bureau (1945–1958), minister of defense (1958–1960), minister of education (1960–1961), president of the Union Générale des Pétroles (1962–1965), president of EDF (1964–1966); Roger Gaspard, general director of EDF (1946–1962), general director of Forges et Ateliers du Creusot (1964–1969), president of the French committee to the World Energy Conference (1970–1974), director of the Bank Crédit Lyonnais (1966–1973); Pierre Massé, associate general director of EDF (1948–1958), minister of planning (1959–1966), president of EDF (1965–1968). Paul Delouvrier, member of different cabinets of ministers of finance in the Fourth Republic, started his career as a military general, then became Delegué Général in Algeria (1958–1960), director of the greater Paris region and prefect (–1969), administrator of the state broadcasting system ORTF (1970–1972), and has been president of EDF since 1969. Below this highest political level the circulation of personnel is a marked phenomenon between the heads of divisions of EDF and the secretariat of the energy commission in the ministry of planning: Lucien Gouni, present director of the EDF division of general economic studies and Albert Robin from the general directorate of EDF had important positions in the elaboration of the last five-year plans. (Information from *Who's Who* and newspapers.)

23. Interview with an official of the Atomforum.

24. Quote from a press conference with the director general of EDF, in P. Simonnot, *Les Nucléocrates* (Grenoble: PUG, 1978), p. 176.

25. Interview with a journalist of *Die Zeit*.

26. Jungk, *Der Atomstaat*, pp. 44 ff.

27. Simonnot, *Les Nucléocrates*, p. 129.

28. Y. Lenoir, *Technocratie française* (Paris: Pauvert, 1977), p. 95.

Chapter 3

1. A. Peyrefitte, *Le Mal français* (Paris: Plon, 1976).

2. For discussion of these widely analyzed relationships, see E. N. Suleiman, *Politics, Power and Bureaucracy in France* (Princeton: Princeton University Press, 1974); A. Darbel and D. Schnapper, *Les Agents du système administratif* (Paris: Mouton, 1969); S. Berger, *The French Political System* (New York: Random House, 1974); H. Ehrmann, *Politics in France* (Boston: Little Brown, 1968); S. Hoffman et al., *In Search of France* (Cambridge: Harvard University Press, 1963); and E. N. Suleiman, *Elites in French Society: The Politics of Survival* (Princeton: Princeton University Press, 1979).

3. See S. Tarrow, *From Center to Periphery* (New Haven: Yale University Press, 1976); and T. C. Thoenig, *L'Ere des technocrates* (Paris: Editions d'Organisation, 1973).

4. The variations in forecasts of energy consumption in the Federal Republic were a major target for criticism. Between 1974 and 1976 such forecasts for the year 2000 went from 400 million tons of coal equivalent (by an independent planning research institute) to 620 million tons (BMFT) and 1,350 million tons (CEE commission). The downward adjustments are often given less publicity than the figures announced by the utilities and nuclear plant supplier firms. See M. Pollak, *Public Involvement in R & D Policies in Germany* (unpublished paper prepared for OECD, 1977), pp. 34 ff.

5. J. P. Colson, *Le Nucléaire sans les français* (Paris: Maspéro, 1977), pp. 114 ff.

6. EDF systematically analyzes the objections presented in DUP procedures. These internal studies are not published. During interviews EDF officials showed these data to the authors.

7. Colson, *Le Nucléaire*, pp. 130 ff.

8. For a description of these procedural irregularities, see *La Gazette nucléaire*, 17 (1978), pp. 7 ff.; Coordination des comités de défense de la Basse Loire (ed.), *Le Pellerin, Cheix-en-Retz* (Chateaurenard: Presses de Chateaurenard, 1978), pp. 15 ff.

9. Interview with an official of EDF.

10. St. Nagel and K. von Moltke, *Citizen Participation in Planning Decisions of Public Authorities, National Report for Germany* (final report prepared for the Environment and Consumer Protection Service, Division G, Commission of the European Community, 1978), p. 35.

11. R. Schäfer, *Praxis des derzeitig gültigen Genehmigungsverfahrens* (unpublished paper presented to a hearing organized by the FDP of Lower Saxony in Wolfenbüttel, on February 7–8, 1977).

12. H. H. Wüstenhagen, *Bürger gegen Kernkraftwerke, Wyhl—der Anfang?* (Reinbek: Rowohlt, 1975), p. 61.

13. Peyrefitte, *Le Mal français*, p. 279.

Chapter 4

1. See discussion in R. Macridis, *French Politics in Transition* (Cambridge, Mass.: Winthrop Publishers, 1975).

2. Schloesing Report, annex no. 23 to the report prepared by the Finance Committee of the National Assembly. Assemblée Nationale, no. 3131, Première session ordinaire de 1977–78, Annexe au procès verbal du la séance du 5 Octobre 1977.

3. L. G. Edinger, *Politics in Germany* (Boston: Little Brown, 1968), p. 255.

4. *Der Spiegel,* 51 (1978), pp. 36 ff.

5. SPD, *Energie—Ein Diskussionsleitfaden* (Bonn: 1977), p. 3.

6. This anecdote was told by an SPD parliamentarian in the discussions of an SPD forum in 1977.

7. *Deutscher Bundestag,* 7, 13871 (1975). See F. Haenschke, *Modell Deutschland? Die Bundesrepublik in der technologischen Krise* (Reinbek: Rowohlt, 1977), pp. 101 ff.

8. H. Gruhl, *Ein Planet wird geplündert* (Frankfurt: Fischer, 1973).

9. Haenschke, *Modell Deutschland,* pp. 90–91.

10. K. Sontheimer, *Grundzüge des Politischen Systems der Bundesrepublik Deutschland* (Munich: 1971); T. Ellwein, *Das Regierungssystem der BRD,* 4d ed. (Cologne: WDV, 1977).

11. RPR, *Le RPR propose. L'Énergie* (Paris: 1977), pp. 13 ff.

12. CDS, *L'Autre solution* (Paris, 1977), pp. 111–112.

13. Ph. St. Marc, *Progrès ou déclin de l'homme* (Paris: Stock, 1978). He expresses a typical view of the ecological position based on Christian humanism.

14. H. W. Ehrmann, *Politics in France,* 3d ed. (Boston: Little Brown, 1976), p. 239.

15. V. Labeyrie, Editorial, *L'Humanité* (August 1, 1977).

16. Parti Socialiste, *Pour une autre politique nucléaire, rapport du comité nucléaire, environnement et société au parti socialiste* (Paris: Flammarion, 1978).

17. K. Winnacker, K. Wirtz, *Das unverstandene Wunder. Kernenergie in Deutschland* (Düsseldorf: Econ, 1975), pp. 76 ff. C. Deubner, *Die Atompolitik der westdeutschen Industrie und die Gründung von Euratom* (Frankfurt: Campus, 1977), pp. 114 ff.

18. See *Frankfurter Allgemeine Zeitung* (January 5, 1978).

19. SPD, *Fachtagung "Energie, Beschäftigung, Lebensqualität"* (Cologne: April 1977), pp. 16 ff.

20. Internal information paper of the SPD youth organization: JUSO-Hessen Süd, *Info* (Frankfurt: February 1977).

21. J. Hallerbach, *Die eigentliche Kernspaltung. Gewerkschaften und Bürgerinitiativen im Streit um die Atomkraft* (Darmstadt-Neuwied: Luchterhand, 1978), pp. 57 ff.

22. This tranformation took place at the end of the 1950s to enable the party to compete with the catch-all CDU and to gain power in Bonn. In the Bad-Godesberg program of 1959, the party adjusted to the political framework of the Federal Republic and abandoned its objectives of deep structural change.

23. See A. Laurens and T. Pfister, *Les Nouveaux communistes* (Paris: Stock, 1973); and A. Harris and A. de Sédouy, *Voyage à l'interieur du parti communiste* (Paris: Le Seuil, 1974). Both books underline the decline in dogmatism and sectarianism, and the growing independence from the Eastern European countries.

24. Present party leader François Mitterand made his political career outside the socialist movement, coming from the republican clubs of the 1960s.

25. This statement must be qualified; one can characterize the French Socialist party as skeptical toward nuclear energy and the Communist party as supportive, but in both contradictory views exist. Engineers from EDF's management are an important group in the PS. In the PCF many intellectuals are close to the environ-

mental movement. See the interview with the socialist technology spokesmen J. M. Belorgey in *Le Sauvage* (March 1978), p. 5. In 1976 the Communist party presented one of its ecology spokesmen, V. Labeyrie, a professor of biology, as a candidate in the partial elections in Tours and symbolically chose green as the campaign color. (See the interview in Les Amis de la Terre periodical *La Baleine* May 1976, pp. 3–5.) In the general elections of May 1978 the CP decided not to present candidates in two electoral districts and back the left-ecologist candidates presented by the *Front Autogestionnaire*, C. Bourdet and S. Depaquit.

26. H. Krasutzky, CGT leader, in a speech on July 19, 1977.

27. After the title of W. I. Lenin's famous polemical article, "Left Radicalism, the Infantile Disorder."

28. CFDT, *Syndicat de l'energie atomique,* l'electronucléaire en France (Paris: Le Seuil, 1975), p. 434.

29. DGB, *Kernenergie im Dienste der Menschheit* (Düsseldorf: 1957).

30. DGB, *Kernenergie und Umweltschutz,* Stellungnahme des DGB Bundesvorstandes (April 5, 1977).

31. For the position of different branch unions, see B. Klaus and K. T. Stiller, *Atomenergie und Gewerkschaftspolitik* (Bielefeld: AJZ-Verlag, 1979).

32. See the documentation of this case in L. Mez and M. Wilke, *Der Atomfilz. Gewerkschaften und Atomkraft* (Berlin: Olle und Wolter, 1977), pp. 195 ff.

33. See Karl D. Bracher, "Problems of Parliamentary Democracy in Europe," in S. R. Graubard (ed.), *A New Europe?* (Boston: Beacon Press, 1963), p. 251; C. Offe, *Strukturprobleme des kapitalistischen Staates* (Frankfurt: Suhrkamp, 1972). For a detailed analysis of all dimensions of this crisis, see J. Habermas, *Legitimation Crisis* (Boston: Beacon Press, 1975).

Chapter 5

1. C. M. Vadrot, *L'Écologie. L'Histoire d'une subversion* (Paris: Syros, 1978), pp. 30 ff.

2. See for example, *Le Monde* (February 11, 1975), p. 32.

3. J. F. Picard and J. M. Fourgous, *La Grande presse dans le débat nucléaire* (Grenoble: IREP, 1977, mimeographed report). This survey includes twenty-one major French newspapers: eleven daily newspapers based in Paris; five daily newspapers based in the provinces and five weekly magazines.

4. *Le Monde* (May 5, 1975).

5. See "Bürgerinitiative." *Zeitung für Kritische Demokraten,* 5 (1977), p. 22.

6. H. H. Wüstenhagen, *Bürger gegen Kernkraftwerke. Wyhl—der Anfang?* (Reinbek: Rowohlt, 1975), pp. 13 ff.

7. The description of site protests is based on interviews with activists and analysis of newspapers and ecologists' publications. For Wyhl see H. H. Wüstenhagen, *Bürger*; B. Nössler and M. de Witt (ed.), *Wyhl. Kein Kernkraftwerk in Wyhl und auch sonst nirgends* (Freiburg: Inform, 1976); Battelle Institut, *Bürgerinitiativen im Bereich von Kernkraftwerken* (Bonn: BMFT, 1975); N. Gladitz (ed.), *Lieber heute aktiv als morgen radioaktiv* (Berlin: Wagenbach, 1976); Institut für marxistische Studien und Forschungen (ed.), *Wyhl—Analyse einer Bürgerbewegung gegen Kernkraftwerke* (Frankfurt: 1976, mimeographed).

8. The Clamshell Alliance has based its tactics on the Wyhl demonstration. Films of the German demonstration are widely disseminated in the United States.

9. See BUU (Bürgerinitiative Umweltschutz Unterelbe) (ed.), *Brokdorf. Der Bauplatz muss wieder zur Wiese werden!* (Hamburg, Association, 1977); Ermittlungsausschuss der BUU (ed.), *Augenzengenberichte aus Brokdorf* (Hamburg: 1976); G. Zint and C. Lutterbeck, *Atomkraft* (Fischerhude: Verlag Atelier im Bauernhaus, 1977).

10. BUU, *Brokdorf,* p. 52.

11. *Frankfurter Rundschau* (January 26, 1977).

12. *Hamburger Abendblatt* (January 19, 1977).

13. Bürgerinitiativen Hameln, Hannover, Göttingen, Kassel, and Ermittlungsausschüsse Hamburg, Kassel (eds.), *Grohnde. Eine Dokumentation* (Hannover: Internationalismus, 1977).

14. Ermittlungsausschuss der nordrhein-westfälischen Bürgerinitiativen (ed.), *Kalkar. Wir das Volk sind nicht gewillt, den Atomtod widerstandslos hinzunehmen* (Cologne: Henke, 1977).

15. F. Fagnani and A. Nicolon (eds.), *Nucléopolis. Matériaux pour l'analyse d'une société nucléaire* (Grenoble: PUG, 1979), pp. 223 ff.

16. *Le Monde* (August 14, 1975). See discussion in Fagnani and Nicolon, *Nucléopolis,* pp. 299–302, for the feeling of impotence among local elected officials. For a study of the mayor's role in France, see M. Kesselman, *French Local Government: The Politics of Consensus* (New York: Alfred Knopf, 1966).

17. D. Anger, *Chronique d'une lutte. Le Combat anti-nucléaire à Flamanville et dans La Hague* (Paris: Simoën, 1977).

18. Th. Jund, *Le Nucléaire contre l'Alsace* (Paris: Syros, 1977).

19. The letter claiming responsibility was signed "Commando Puig Antich-Ulrike Meinhof." This name suggests relationships with the German Red Army fraction. Nevertheless, the authors of the crime were never identified.

20. See different reactions to this act of terrorism in Vadrot, *L'Écologie,* pp. 82–83.

21. Collectif d'Enquête, *Aujourd'hui Malville—demain la France!* (Saint Etienne: La Pensée Sauvage, 1978). Conseil Général de l'Isère, *Creys-Malville. Le Dernier mot?* (Grenoble: PUG, 1977).

22. Gladitz (ed.), *Lieber heute,* p. 83.

23. Ulrich Freund argues that the right of life is part of the constitutional order and that individuals threatened by a nuclear power plant have a right to resist the technology. U. Freund, "Widerstandsrecht gegen Kernkraftwerke?" in *BBU-Aktuell,* 1 (1978), pp. 27 ff.

Chapter 6

1. The letter claiming the bombing at Fessenheim was signed "Commando Puig Antich-Ulrike Meinhof."

2. Press release after the bombing at Fessenheim "Les Amis de la Terre—Paris."

3. Umweltwissenschaftliches Institut des BBU (ed.), *Aktionskatalog des BBU* (Stuttgart: Fantasia, 1978), p. 4.

4. *Junge Kirche,* 1 (1977), special supplement: "Von Wyhl nach Brokdorf. Hat die gewaltfreie Aktion in der Bewegung gegen Atomkraftwerke noch eine Chance?" p. 5.

5. Ibid., p. 6.

6. *BBU-Aktuell*, 1 (1978), p. 39.

7. *Alternatives*, 5 (1978), special issue: "Désobéissance civile et luttes autonomes," p. 137.

8. Umweltwissenschaftliches Institut des BBU (ed.), *Aktionskatalog*. See also H. C. Buchholtz et al. (ed.), *Widerstand gegen Atomkraftwerke. Informationen für Atomkraftwerkgegner und solche, die es werden wollen* (Wuppertal: Hammer, 1978); Aktionsgemeinschaft für Umweltschutz Darmstadt (ed.), *KKW—Fibel für Bürgerinitiativen* (Hamburg: Verlag für das Studium der Arbeiterbewegung, 1977).

9. L. Samuel, *Guide pratique de l'écologiste* (Paris: Belfond, 1978).

10. Umweltwissenschaftliches Institut des BBU (ed.), *Aktionskatalog*, p. 4.

11. See "Dossiers et documents, les élections legislatives de Mars 1978," *Le Monde* (Paris, 1978).

12. *Der Spiegel*, 24 (1978), pp. 30 ff.

13. *La Gazette nucléaire*, 24 (1979), p. 10.

14. M. Lambert (ed.), *Nucléaire—Peur sur La Hague* (Coutances: Ocep, 1977). See also the occasional publication: *Le Petit cafard des falaises. Journal de lutte Flamanville—La Hague.*

15. *La Gazette nucléaire*, 24 (1979), p. 12; *La Gazette nucléaire*, 4 (1976), p. 3.

16. *La Gazette nucléaire*, 12 (1977), p. 4 ff.

17. S. Rippon, "The European Reprocessing Scene: Political, Economic and Technical" (paper presented at the International Conference on the Nuclear Fuel Cycle, London, September 26–29, 1978), p. 9.

18. *Le Canard enchaîné*, March 7, 1979, p. 2.

19. *Der Spiegel*, 54 (1977), p. 54.

Chapter 7

1. *Der Spiegel*, 50 (1978).

2. Appel des 400, in *Le Monde* (February 11, 1975).

3. Groupe de Bellerive, Projet Alter, *Esquisse d'un régime à long terme tout solaire* (Paris: Syros, 1978, mimeographed).

4. CFDT (ed.), *L'Electronucléaire en France* (Paris: Le Seuil, 1975); CFDT (ed.), *Les Dégâts du progrès* (Paris: Le Seuil, 1977), pp. 186 ff.

5. See IEJE (ed.), *Réflexions sur les choix energétiques français* (Grenoble, 1975, mimeographed); IEJE (ed.), *Alternatives au nucléaire* (Grenoble, 1975, mimeographed).

6. See K. Winnacker and K. Wirtz, *Das unverstandene Wunder, Kernenergie in Deutschland* (Düsseldorf: Econ, 1975), pp. 124–125.

7. *Bundestag* Proceedings: Session 215 (January 22, 1976), p. 14935.

8. Autorengruppe des Projektes SAIU an der Universität Bremen, *66 Erwiderungen zum besseren Verständnis der Kernenergie* (Berlin: Oberbaumverlag, 1975).

9. Arbeitsgruppe Wiederaufarbeitung an der Universität Bremen, *Atommüll oder der Abschied von einem teuren Traum* (Reinbek: Rowohlt, 1977).

10. Tutorium Umweltschutz, *Zur Notwendigkeit eines Moratoriums. Aufforderung zu einer öffentlichen Diskussion* (Heidelberg: 1975, mimeographed).

11. For the relationships between career strategies and scientific age, see P. Bourdieu, "The Specificity of the Scientific Field and the Social Conditions of the Progress of Reason," in *Social Sciences Information*, 14, 6 (1975), pp. 19 ff. Helga Nowotny made a detailed empirical study of the different strategies developed by pro- and antinuclear scientists in the public controversy in Austria; see H. Nowotny, *Kernenergie: Gefahr oder Notwendigkeit. Anatomie eines Konflikts* (Frankfurt: Suhrkamp, 1979).

12. A typical example of this argument is given in E. Münch and P. Borsch, "Gibt es eine wissenschaftliche Kernenergiekontroverse," in *Atomwirtschaft* (January 1977), pp. 27 ff.

13. *Le Monde* (February 11, 1975), p. 32.

14. In late 1975 scientists organized both antinuclear (Tutorium Umweltschutz, *Zur Notwendigkeit*) and pronuclear petitions (Offener Brief an die Abgeordneten des deutschen Bundestags, November–December 1975).

15. *Le Monde* (March 19, 1974).

16. *Niedersächsisches Ärzteblatt*, 14 (1974, *Gesundheitliche Gefahren durch Atomkraftwerke*); 17 (1974, *Gewässergüte der Weser nicht verschlechtern*); 18 (1974, *Kernkraftwerke und Umwelt*).

17. *Der Spiegel*, 32 (1978), pp. 43 ff.

18. *La Gazette nucléaire*, 5 (1977).

19. H. Strohm (ed.), *Atomenergie und Arbeitsplätze* (Hamburg: Association, 1978); Gottlieb Duttweiler Institut (ed.), *Kernenergie–offen bilanziert* (Frankfurt: Fischer, 1976); B. Manstein (ed.), *Atomares Dilemma* (Frankfurt: Fischer, 1977), pp. 124 ff.; Bürgerinitiative Chemiekollegen gegen AKW, *Atomenergie und Arbeitsplätze* (Hamburg: Hein, 1978).

20. The *Revue de Protection contre les Rayonnements Ionisants* (1978) refers to the year 1978 as the 83rd year of radiation and the 36th year of atomic chaos.

21. Louis Puiseux in an interview: *Le Nouvel Observateur* (April 17, 1978).

22. See, for example, the controversy in the periodical of the French consumers' association: *Que choisir?*, special issue: "Nucléaire. Le face à face" (1977).

23. Leaflet issued by social scientists in Paris, *Nucléaire–Choix de société* (May 1978).

24. Manstein (ed.), *Atomares*, p. 64.

25. Ibid., p. 114.

26. H. Strohm, *Friedlich in die Katastrophe. Eine Dokumentation über Kernkraftwerke* (Hamburg: Association, 1977).

27. *Que choisir?*, special issue: "Au Soleil de l'an 2000. Peut-on stopper le nucléaire?" (1978).

28. *Der Spiegel*, 26 (1977).

29. CNRS Groupe de Travail, "L'Energie Nucléaire," *Le Courriere du CNRS*, 19 (January 1976), pp. 23–26.

30. Th. Jund, *Le Nucléaire contre l'Alsace* (Paris: Syros, 1977), p. 72.

31. *Que choisir?* "Au Soleil."

32. L. Le Prince-Ringuet, "Atoms, Science and Politics," *Development Forum*, 3 (June 1975), pp. 9 ff.

33. See, for example, H. Hülsmann and R. Tschiedel (eds.), *Kernenergie und wissenschaftliche Verantwortung* (Kronberg/Ts: Athenäum, 1977); H. H. Wüstenhagen, K. Krusewitz, H. J. Krysmanski, R. Krysmanski, and M. O. Hinz, *Umweltmisere, Bürgerinitiativen und die Verantwortung der Wissenschaftler* (Cologne: Pahl-Rugenstein, 1976).

34. GSIEN position paper published in *Que choisir?* (1977).

35. Appel des 400, in *Le Monde*.

Chapter 8

1. C. Tilly et al., *The Rebellious Century* (Cambridge, Mass.: Harvard University Press, 1975), p. 23.

2. See O. Rammstedt, *Soziale Bewegung* (Frankfurt: Suhrkamp, 1978).

3. Numerous works in social history confirmed this general hypothesis for different periods and movements; see, for example: Tilly et al., *The Rebellious Century*; E. Hobsbawm, *Primitive Rebels* (New York: Norton, 1965); Rammstedt, *Soziale*; G. Botz, G. Brandstetter, and M. Pollak, *Im Schatten der Arbeiterbewegung. Zur Geschichte des Anarchismus in Österreich und Deutschland* (Vienna: Europa, 1977).

4. For statistical data on changes in economic structure and urbanization and migration patterns in France, see I. Chiva, "Les Changements et tendances manifestés durant les vingt dernières années dans la société française," in *Colloque sur les implications psychosociologiques du développement de l'industrie nucléaire, 13–14 Janvier 1977* (Paris: Société Française de Radioprotection, 1977), pp. 16 ff.

5. D. Menyesch and H. Unterwedde, "Wirtschaftliche und soziale Strukturen in der Bundesrepublik und in Frankreich," in R. Picht (ed.), *Deutschland, Frankreich, Europa. Bilanz einer schwierigen Partnerschaft* (Munich: Piper, 1978), p. 53.

6. Chiva, "Les Changements."

7. *Le Nouvel Observateur* (August 1978).

8. B. Schäfers, *Sozialstruktur und Wandel der Bundesrepublik Deutschland* (Stuttgart: DTV, 1976), pp. 242 ff.

9. R. Inglehart, *The Silent Revolution: Changing Values and Political Styles among Western Publics* (Princeton: Princeton University Press, 1977).

10. A. W. Gouldner, *The Future of Intellectuals and the Rise of the New Class* (New York: Continuum Books, The Seabury Press, 1979).

11. Unfortunately survey institutes have not posed directly comparable questions about nuclear power until recently. To compare survey findings over a longer period, and in particular to show the changes after 1974, we must use different sources: in France an IFOP survey in 1974 and the SOFRES-*Figaro* survey in 1978; in Germany a survey from the Sample Institute of Hamburg in 1975 and from INFAS in 1977.

12. The term "professionals" in continental Europe is used differently than in the American context. "Professionals" in German *(freie Berufe)* and in French *(professions libérales et cadres supérieurs)* refer to people in positions that depend on fees rather than salaries (lawyers, physicians, dentists, architects). They are listed in different statistical categories. In France they are usually classified with managers of large industry because of income similarities, whereas in Germany they are not aligned with any other income group.

13. The persistent opposition of the Communist party against the government's nuclear policy suggests, however, that CP sympathizers are more sympathetic to the party line than this survey result indicates.

14. In part at least, this difference might result from different questions asked in the surveys.

15. *Le Nouvel Observateur* (August 1977).

16. A. Touraine, *The May Movement* (New York: Random House, 1971).

17. S. Hoffman (ed.), *In Search of France* (Cambridge, Mass.: Harvard University Press, 1963).

18. Menyesch and Unterwedder, "Wirtschaftliche," p. 79.

19. Ibid., p. 79.

20. A. Buro, "Historische Erfahrungen und ausserparlamentarische Politik," in J. Hallerbach (ed.), *Die eigentliche Kernspaltung. Gewerkschaften und Bürgerinitiativen im Streit um die Atomkraft* (Darmstadt-Neuwied: Luchterhand, 1978), p. 20.

21. Ibid., p. 22.

22. K. A. Otto, *Vom Ostermarsch zur APO. Geschichte der ausserparlamentarischen Opposition in der Bundesrepublik, 1960–1970* (Frankfurt: 1977).

23. S. Berger, "The French Political System," in S. Beer (ed.), *Patterns of Government,* 3d ed. (New York: Random House, 1973), p. 340.

Chapter 9

1. P. Delmon et al., *La Participation des français à l'amélioration de leur cadre de vie* (Paris: 1976, mimeographed report).

2. Ministère de la Culture et de l'Environnement, Service de l'Information des Relations et de l'Action Educative, *Les Associations et l'environnement, liste indicative des principales associations nationales et régionales* (Paris: 1978, mimeographed report).

3. A. Rose, "Voluntary Associations in France," in A. Rose (ed.), *Theory and Method in the Social Sciences* (Minneapolis: University of Minnesota Press, 1954), pp. 80 ff.

4. M. Crozier, *The Stalled Society* (New York: Viking Press, 1973), pp. 67–68.

5. H. W. Ehrmann, *Politics in France,* 3d ed. (Boston: Little Brown), p. 57. Besides the Catholic church, the Protestant church and the Masons also created voluntary associations, but these associations are small and had little influence on the emerging environmental groups of the 1960s.

6. See L. Boltanski, "Taxinomies sociales et luttes de classes," in *Actes de la recherche en sciences sociales,* 29 (1979), pp. 75–105; E. Poulat, *Eglise contre bourgeoisie* (Paris: Casterman, 1977). See also R. O. Paxton, *Vichy France* (New York: Knopf, 1972); S. Hoffman, *Essais sur la France. Echec ou renouveau* (Paris: Le Seuil, 1974), pp. 17 ff.

7. In 1976 the German federal environmental agency listed 73 national organizations and 590 in the different *Länder*; see Umweltbundesamt, Nichtstaatliche Umweltschutz-Organisationen in der BRD, Materialien 1/75 (1976, mimeographed).

8. The citizen initiatives are the most significant new phenomenon in German political life of the 1970s. In the last few years more than 300 books and articles have been published on the citizen initiatives. See the bibliography put together by B.

Welz, in B. Guggenberger and V. Kempf (eds.), *Bürgerinitiativen und repräsentatives System* (Opladen: WDV, 1978), pp. 375 ff.

9. U. Thaysen, "Bürgerinitiativen, Parlamente, und Parteien in der Bundesrepublik. Eine Zwischenbilanz," in *Zeitschrift für Parlamentsfragen,* 1 (1978), p. 90. See also Emnid Information (November 11–12, 1973), p. 7.

10. Surveys in 1972 found that 43 percent of the initiatives in Bavaria and 45 percent of those in the Ruhr were organized around environmental problems. In Berlin the focus was rather on housing and urban development problems. See Bayerisches Staatsministerium des Inneren, *Bürgerinitiativen in Bayern* (Munich: 1973, Bestandsaufnahme Az I Bl-3000-72/1); B. Borsdorf-Ruhl, *Bürgerinitiativen im Ruhrgebiet* (Essen: 1973). In Germany as a whole the initiatives were organized as follows: environment 16.9 percent, transportation 11.8 percent, kindergartens and sports facilities 15.8 percent, schools 8.1 percent, urban development 8 percent, marginal groups 7.1 percent, cultural life 5.7 percent, youth problems 4.9 percent, community facilities 3.9 percent, urban sanitation 3.6 percent, and cultural activities 3.3 percent. See P. von Kondolitsch, "Gemeindeverwaltungen und Bürgerinitiativen," in *Archiv für Kommunalwissenschaften,* 14 (1975), pp. 264 ff.

11. J. Carlier, *Vanoise. Victoire pour demain* (Paris: Calmann-Lévy, 1972).

12. Interview with S. Moscovici in J. P. Ribes (ed.), *Pourquoi les écologistes font-ils de la politique?* (Paris: Le Seuil, 1978), p. 138.

13. B. Borsdorf-Ruhl, *Bürgerinitiativen,* pp. 78 ff.

14. See L. Gerlach, "Energy Wars and Social Change," delivered at 1978 meeting of Southern Anthropological Society, April 12–15, 1978 (mimeographed); and L. Gerlach and V. Hine, *Lifeway Leap* (Minneapolis: University of Minnesota Press, 1973), pp. 163 ff.

15. He based the issue on a panel discussion of internationally known intellectuals and politicians, including Herbert Marcuse, Edgar Morin, Sicco Mansholt, and René Dumont.

Chapter 10

1. We use the term ideology neither in a restricted Marxist sense as distorted or selected ideas in defense of particular interests nor in a neopositivist sense which maintains the distinction between a scientific point of view and the commonly used lay interpretations of reality. By analyzing the antinuclear ideology, we start from the fundamental contradiction established by linguistic analysis that discourse is not the same as reality and that political discourse is intended to persuade and does not necessarily seek analytic clarity. J. L. Austin, *How to Do Things with Words* (Cambridge, Mass.: Harvard University Press, 1962). As a strategic tool, the ecological discourse first constructs a common scheme of reference as a precondition for intentional action. See A. V. Cicourel, *Cognitive Sociology: Language and Meaning in Social Interaction* (Middlesex: Penguin, 1973), p. 34.

2. Interview with S. Moscovici in J. P. Ribes (ed.), *Pourquoi les écologistes font-ils de la politique?* (Paris: Le Seuil, 1978), pp. 105–106.

3. J. Dahl, "Kommt Zeit kommt Unrat," in *Scheidewege,* 1 (1977), p. 39.

4. See *Arbeiterkampf. Zeitung des kommunistischen Bundes* (Hamburg, 1977, special issue: "Warum kämpfen wir gegen Atomkraftwerke?"), pp. 19 ff.

5. C. Guedeney and G. Mendel, *L'Angoisse atomique et les centrales nucléaires* (Paris: Payot, 1973), pp. 61 ff.

6. "KKW—Tod auf Raten" was the slogan on a German antinuclear poster and bumper sticker.

7. This was written during the Wyhl site occupation by K. F. E. Weisgräber.

8. *Natur und Umwelt*, 3 (1976), p. 6.

9. See W. Laqueur, *Weimar: A Cultural History, 1918–1933* (New York: G. P. Putnam, 1974).

10. E. Kuhl, "Wie ist es zur Ölkrise gekommen und wie antworten wir darauf?" in *Natur und Umwelt*, 1 (1976), pp. 8–9.

11. See H. Winock, *Histoire politique de la revue "Esprit," 1930–1950* (Paris: Le Seuil, 1975).

12. P. St. Marc, *Progrès ou déclin de l'homme?* (Paris: Stock, 1978), p. 118.

13. H. M. Enzensberger, "Zur Kritik der politischen Ökologie," in *Kursbuch*, 33 (1973), p. 5.

14. Arthur in *La Gueule ouverte* (March 8, 1977).

15. A. Touraine identifies such communal utopias as typical for the pre-political stage of social movements. Protest movements often justify their claims by referring to traditional cultural values and by elaborating communal utopias opposed to the state ideology. In Touraine's view, however, social movements could bring about change only once they define themselves in terms of class relationships rather than in terms of lifestyles. A. Touraine, *La Voix et le regard* (Paris: Le Seuil, 1978), pp. 175–177.

16. Front Autonomiste Socialiste-Autogestionnaire Breton, in *Combat Breton* (May 1976).

17. A. M. Vilaine, "Ecoutez-moi," in *Le Sauvage*, 1 (1978), p. 30.

18. One has to keep in mind that the ecology movement has emerged at the same time as various apolitical rural communities and the spread of vegetarianism and holistic health groups. See D. Leger, "Les Utopies du 'retour'," in *Actes de la recherche en sciences sociales*, 29 (1979), pp. 54 ff.

19. Urban Indians consider themselves part of the spontaneous unorthodox left. By wearing Indian dress, they symbolically identify with marginal groups.

20. L. Puiseux, in an interview for *La Gueule ouverte*, cited in A. Gorz and M. Bosquet, *Ecologie et politique* (Paris: Le Seuil, 1975), pp. 122–123.

21. Nature et Progrès (ed.), *L'Ecologie intégrale* (tract, no year), p. 2.

22. *Blauer Brief* (October 1977, newsletter for the members of the Bavarian Catholic agricultural youth organization).

23. *Bürgerinitiative*, 5 (1976), p. 5.

24. H. Weinzierl, "Natur als Alternative," in *Natur und Umwelt*, 3 (1976), p. 1.

25. H. Gruhl in his speech at Brokdorf.

26. See J. L. Parodi, "Le Mouvement écologique dans le système des partis français. Essai de problématique," in *Revue politique et parlementaire*, 878 (1979), p. 19.

27. See Laqueur, *Weimar*; K. Sontheimer, *Antidemokratisches Denken in der Weimarer*

Republik. Die politischen Ideen des deutschen Nationalismus zwischen 1918 und 1933, (Munich: Nymphenburger Verlag, 1962).

28. P. St. Marc, *Progrès,* p. 387.

29. Interview with Brice Lalonde, in Ribes (ed.), *Pourquoi,* p. 29.

30. See Programm der AUD, pp. 18–20; Das Grüne Manifest. Programm der Partei Grüne Aktion Zukunft.

31. P. St. Marc, *Progrès,* p. 239.

32. I. Illich, *Tools for Conviviality* (New York: Harper and Row, 1973).

33. ADRET (ed.), *Travailler deux heures par jour* (Paris: Le Seuil, 1977).

34. Interview with H. Laborit in *Le Sauvage,* 1 (1978), pp. 24 ff.

35. H. Kampinski, "Ursachen und Auswirkungen des Fehlverhaltens der heutigen Menschheit zu ihrer Umwelt und zum Planeten Erde," in *Natur und Umwelt,* 1 (1976), p. 5.

36. *Natur und Umwelt,* 3 (1976), p. 6.

37. Kuhl, *Wie ist es,* p. 9.

38. W. Harich, *Kommunismus ohne Wachstum* (Reinbek: Rowohlt, 1975).

39. See the classical text O. Spengler, *Preussentum und Sozialismus* (Munich: Beck, 1922).

40. Lambert, "Les Confessions d'un maniaque," in *La Gueule ouverte,* 197 (1978), p. 5.

41. *Légitime Défense,* Document of Les Amis de la Terre (Lille), 13 (1978), p. 56.

Chapter 11

1. J. P. Colson, *Le Nucléaire sans les français* (Paris: Maspéro, 1978), p. 139.

2. E. Gaul, "Die Fragwürdigkeit des Rechts—Am Beispiel der Atomenergie," in B. Manstein (ed.), *Atomares Dilemma* (Frankfurt: Fischer, 1977), p. 69.

3. See A. T. von Mehren and J. Grodly, *The Civil Law System,* 2d ed. (Boston: Little Brown, 1977), pp. 97–137; E. Rehbinder, *Right of Associations and Other Legal Persons to Sue in Cases Concerned with the Protection of the Environment* (Commission of the European Community, ENV 738/77, 1977), pp. 3–156.

4. U. Freund, "Widerstandsrecht gegen Kernkraftwerke," in *BBU-Aktuell,* 2 (1978), pp. 27–34.

5. See remarks by J. Lecanuet in *Revue juridique de l'environnement,* 3–4 (1976), pp. 10 ff.

6. See Rehbinder, *Right of Associations* (p. 10 for Germany, p. 91 for France) and OECD *Nuclear Law Bulletin,* 15 (April 1975).

7. In France the maximum allowable annual dosage is 500 millirems (OECD, *Nuclear Law Bulletin,* p. 149). In Germany the new amendments to the atom law reduced the maximum allowable dosage from 150 to 30 millirems. However, in the event of an accident the maximum safe dosage for those in the vicinity is 5 rem (*Nuclear Law Bulletin,* 18, December 1976, p. 24). See also OECD, Nuclear Energy Agency, *Nuclear Legislation* (Paris: OECD, 1972).

8. Colson, *Le Nucléaire,* p. 142.

9. *Le Matin de Paris* (April 19, 1977). F. Gravelaine and S. O'Dy, in *L'Etat EDF* (Paris: Alain Moreau, 1968), p. 169, argue that prepermit construction can occur wherever prefectural authority is strong. P. Colson *(Le Nucléaire)*, p. 136, says it occurs wherever EDF already owns the land.

10. See L. Samuel, *Guide pratique de l'écologiste* (Paris: Belfond, 1968), pp. 94–95.

11. For discussion of rules of standing, see Rehbinder *Right of Associations,* pp. 6–15 for Germany, pp. 88–94 for France.

12. Art. 40 of the *Law of 10 July 1976 on the Protection of Nature.* Arts. L.121–8 and L.160–1 of the *Urban Code* also allow standing to official registered environmental associations that wish to uphold planning statutes concerning recreation areas, sites, and monuments.

13. K. Boisserée, "Cooperation and Participation in the Protection of the Environment," *Environmental Policy and Law,* 1 (1975), pp. 22–23. Also *Der Spiegel,* 13 (1977), p. 46.

14. J. C. Barreau (ed.), *L'Escroquerie nucléaire* (Paris: Stock, 1978); Rehbinder, *Right of Associations,* p. 154.

15. Sec. 7 of the atom law (Atomgesetz), p. 3057. In France the terminology is more ambiguous: "adequate precautions" must be taken before licensing. See Rehbinder, *Right of Associations,* p. 111.

16. Colson, *Le Nucléaire,* p. 112. Note that to annul a permit may take two years and the proceeding itself does not require suspension of work.

17. From administrative court procedures act, sec. 80, quoted in *Nuclear Law Bulletin,* 15 (April 1975), p. 25.

18. Accounts of the French cases are in Samuel, *Guide pratique,* pp. 82–96, and in Colson, *Le Nucléaire,* pp. 142–144.

19. Henri Fabre-Luce in *La Baleine,* quoted in Samuel, *Guide pratique,* p. 82.

20. *Nuclear Law Bulletin,* 13 (April 1974), pp. 18–20.

21. *Nuclear Law Bulletin,* 15 (April 1975), pp. 24–27.

22. H. Albers, "Möglichkeiten und Grenzen gerichtlicher Uberprüfung anhand der neueren Rechtsprechung auf dem Gebiet des Atomrechts," in H. Hulsemann and R. Tschiedel (eds.), *Kernenergie und wissenschaftliche Verantwortung* (Kronberg/Ts: Athenäum Verlag, 1977), pp. 79–111.

23. *Süddeutsche Zeitung* (October 18, 1977), p. 5.

24. Ibid. (March 15, 1977).

25. Quoted in *Süddeutsche Zeitung* (March 18, 1977). The court also expressed doubts as to whether the West German people had chosen freely to accept the risks of nuclear power or simply accepted the risks because the wheel of progress cannot be reversed.

26. *Süddeutsche Zeitung* (May 31, 1977), p. 14.

27. Colson, *Le Nucléaire,* p. 103.

28. Samuel, *Guide pratique,* p. 84.

29. Interview with a professor of environmental law.

30. E. Rehbinder, "Controlling Environmental Enforcement Deficits: West Germany," *American Journal of Comparative Law,* 24 (1976), p. 384.

31. Interview with an environmental lawyer.

32. See editorial in *Süddeutsche Zeitung* (March 15, 1977), p. 4, for positive reaction of environmentalists and local Wyhl protestors. For criticism, see *Süddeutsche Zeitung* (April 14, 1977).

33. Adolf Schmidt, quoted in *Der Spiegel,* 13 (1977), p. 46.

34. *Nuclear Law Bulletin,* 20 (December 1977), pp. 21–22.

35. *Der Spiegel,* 32 (1977), pp. 45–46.

Chapter 12

1. W. J. Mommsen, "Nationalbewusstsein und Staatsverständnis der Deutschen," in R. Picht, *Deutschland, Frankreich, Europa. Bilanz einer schwierigen Partnerschaft* (Munich: Piper, 1978), p. 35.

2. A. Mitscherlich and M. Mitscherlich, *The Inability to Mourn* (New York: Grove Press, 1975).

3. J. Hayward, *The One and Indivisible French Republic* (London: Weidenfeld and Nicolson, 1973).

4. A. Nicolon, "Analyse d'une opposition à un site nucléaire," in F. Fagnani and A. Nicolon, *Nucléopolis* (Grenoble: PUG, 1979), pp. 244 ff.

5. *Le Monde* (December 1, 1978).

6. See M. Popp and K. Lang, "The Public Discussion about the Peaceful Utilization of Nuclear Energy in the Federal Republic of Germany and the Information and Discussion Campaign of the Federal Government," in IAEA, *International Conference on Nuclear Power and its Fuel Cycle* (Salzburg, May 1977).

7. Ibid., p. 20.

8. "Nuclear Energy—Information for the Citizen," "Discussions and Interviews on Nuclear Energy," "Information Letter on Nuclear Energy."

9. H. Matthöfer, *Interviews und Gespräche zur Kernenergie,* 2d ed. (Karlsruhe: Müller, 1977), pp. 10–11.

10. Popp and Lang, *Public Discussion,* p. 15.

11. Matthöfer, *Interviews,* p. 11.

12. This case illustrates what Theodore J. Lowi has developed in his argument on why liberal governments cannot plan. See T. J. Lowi, *The End of Liberalism* (New York: Norton, 1969), pp. 101 ff.

13. *Le Nouvel Observateur* (August 8, 1977).

14. L. Puiseux, *La Babel nucléaire* (Paris: Editions Galilée, 1977).

15. *Le Nouvel Observateur* (November 21, 1977), p. 104.

16. By 1979 the governmental coalition parties, SPD and FDP, abandoned routine inquiries into the political attitudes of civil service candidates.

17. *Der Spiegel,* 10 (1977), p. 21.

18. *Frankfurter Rundschau* (April 2, 1978).

19. *Le Monde* (June 6, 1979).

20. *Der Spiegel,* 51 (1978), p. 55.

21. *Konkret,* 1 (1978), p. 20.

22. Arbeitsgemeinschaft für Umweltfragen, *Das Umweltgespräch. Bürgerinitiative-Bürgerbeteiligung* (Bonn: AFU, 1977).

23. *Der Spiegel,* 21 (1979), p. 19.

24. They included Amory Lovins, Alice Stewart, Frank Barnaby, and Dean Abrahamson.

25. See, for example, *Frankfurter Rundschau* (April 2, 1979); *Die Zeit,* 15 (1979), pp. 3–5; *Die Zeit,* 16 (1979), pp. 3–4.

26. *Frankfurter Rundschau* (April 4, 1979), p. 3.

Chapter 13

1. Among the other west European nations, the United Kingdom, and to a lesser degree, Sweden, have built up their own independent nuclear capacity; but Great Britain's technological choice of the gas-graphite reactor has restricted her nuclear industry to domestic production.

2. Österreich Dokumentation, *Kernenergie,* vols. 1–4 (Vienna: Bundespressedienst, 1977). H. Hirsch and H. Nowotny, "Information and Opposition in Austria's Nuclear Energy Policy," in *Minerva,* 15, 3–4 (1978), pp. 314–334. D. Nelkin and M. Pollak, "The Politics of Participation and the Nuclear Debate in Sweden, the Netherlands, and Austria," in *Public Policy,* 25, 3 (1977), pp. 333–357.

3. Nelkin and Pollak, "The Politics of Participation."

4. In Denmark a referendum is scheduled in early 1980.

5. L. Mez, "Die Atomindustrie in Westeuropa," in *Technologie und Politik,* 7 (1977), pp. 140–141.

6. *Frankfurter Allgemeine Zeitung* (May 2, 1979).

Chapter 14

1. E. Suleiman, *Elites in French Society* (Princeton: Princeton University Press, 1978), ch. 9.

2. J. Habermas, *Legitimation Crisis* (Boston: Beacon Press, 1975).

3. C. Offe, "The State, Ungovernability and the Search for the Non-Political," Conference on Individual and the State, February 3, 1979, organized by the Canadian Broadcasting Corporation and the University of Toronto. See also P. Rosanvallon and P. Viveret, *Pour une nouvelle culture politique* (Paris: Le Seuil, 1977).

4. J. Straussman, *The Limits of Technocratic Politics* (Rutgers: Transaction Books, 1978).

5. According to a Harris survey commissioned by EDF public support for nuclear power has increased to 57 percent over the last two years, unaffected by the Three Mile Island accident. This could indicate the weakness of the antinuclear movement combined with the deepening economic crisis. A German poll by the Allensbach Institut für Demoskopie finds that public opinion remains ambiguous and that skepticism remains high: 30 percent wanted Germany to continue building nuclear plants; 37 percent felt no more should be built but those in operation should be used; 24 percent wanted to discontinue those in existence as well. However, to a question put in general terms, a majority of 53 percent claimed to be

more in favor than against building new nuclear power plants. Our study suggests, however, that neither short-term oscillations in the strength and activity of the antinuclear movement nor changes in public attitudes allow a linear interpretation of the future of the controversy.

6. A. Touraine, *La Voix et le regard* (Paris: Le Seuil, 1978), pp. 174 ff.

7. L. Puiseux, *La Babel nucléaire* (Paris: Editions Galilée, 1977), p. 167.

8. *Frankfurter Allgemeine Zeitung* (May 2, 1979).

Index